T0137128

Studies in Systems, Decision and Control

Volume 63

Series editor

Janusz Kacprzyk, Polish Academy of Sciences, Warsaw, Poland
e-mail: kacprzyk@ibspan.waw.pl

About this Series

The series "Studies in Systems, Decision and Control" (SSDC) covers both new developments and advances, as well as the state of the art, in the various areas of broadly perceived systems, decision making and control- quickly, up to date and with a high quality. The intent is to cover the theory, applications, and perspectives on the state of the art and future developments relevant to systems, decision making, control, complex processes and related areas, as embedded in the fields of engineering, computer science, physics, economics, social and life sciences, as well as the paradigms and methodologies behind them. The series contains monographs, textbooks, lecture notes and edited volumes in systems, decision making and control spanning the areas of Cyber-Physical Systems, Autonomous Systems, Sensor Networks, Control Systems, Energy Systems, Automotive Systems, Biological Systems, Vehicular Networking and Connected Vehicles, Aerospace Systems, Automation, Manufacturing, Smart Grids, Nonlinear Systems, Power Systems, Robotics, Social Systems, Economic Systems and other. Of particular value to both the contributors and the readership are the short publication timeframe and the world-wide distribution and exposure which enable both a wide and rapid dissemination of research output.

More information about this series at http://www.springer.com/series/13304

Edy Portmann · Matthias Finger
Editors

Towards Cognitive Cities

Advances in Cognitive Computing
and its Application to the Governance
of Large Urban Systems

 Springer

Editors
Edy Portmann
Institute of Information Systems (IWI)
University of Bern
Bern
Switzerland

Matthias Finger
Institute of Technology and Public Policy
École Polytechnique Fédérale de Lausanne
Lausanne
Switzerland

ISSN 2198-4182   ISSN 2198-4190   (electronic)
Studies in Systems, Decision and Control
ISBN 978-3-319-81606-7   ISBN 978-3-319-33798-2   (eBook)
DOI 10.1007/978-3-319-33798-2

Printed on acid-free paper

This Springer imprint is published by Springer Nature
The registered company is Springer International Publishing AG Switzerland

Foreword

Since the beginning of the third millennium, the urban system has become the global population's predominant habitat. In those parts of the world that are mostly urbanized—Europe, the Americas, Australia—we can reflect on the status and the development of the city. In other parts of the world where the urbanization process is underway—Asia and Africa—the making and managing of cities is of highest concern. The expectations of cities vary widely, depending on location, climate, and political system. Yet the contribution of cities for societal progress, sustainability, resilience, and livability becomes as clear as their contribution toward inequality, pollution, and other risks. At the same time, the necessity to build more than 2 billion urban living and working spaces in the coming three decades demonstrates the need for a new basis for planning and managing cities. Therefore, this book will define important new contributions and will ultimately describe what can be achieved by cognitive cities.

The planning of cities is in a deep crisis, and at the same time, new opportunities for city design and management emerge. Ever since the middle of the twentieth century and the deliberations of Jane Jacobs, established planning paradigms have been questioned. In addition, the emergence of rapidly growing urban systems in Asia, Africa, and South America that seemingly can do without established city planning methods has shed a new light on the challenges of and to the city in the twenty-first century.

Observing first hand the rapid growth of urban systems such as Singapore, Jakarta, Yangon, or Shanghai, one cannot fail to observe a tremendous difference to the planning of cities such as Chandigarh or Brasilia in South America around the middle of the last century. Currently, emerging cities such as Shenzhen grow to mega city status within less than 30 years. It is almost inconceivable that this type of growth could have been planned traditionally, especially when compared to the tremendous efforts for planning and implementing even minor expansions in European cities.

What, then, makes these cities emerge? What are the driving forces behind their relentless growth? One reason could be the creation of special economic zones and advantageous geographic locations. Another reason is that small, medium, and large local communities, enabled by the advancement of information technology, reached the capacity to develop their direct vicinity and to connect to neighboring and next-level urban systems. However, this does not explain the explosive urban growth rates. Instead, we observe that a new way of planning and managing cities is emerging.

Cities are made by people for people—hopefully. In order to participate in the creation, transformation, or management of cities, their present or future citizens need to be involved and empowered. Information technology now offers tools to support this development. Citizen design science, cognitive design computing, and the underlying artificial intelligence are three foundations for the interaction of citizens with the cities of the future.

Cognitive design computing extends the application of IT in cities into new areas. Cognitive computing and artificial intelligence, both concepts of the late 1950s, went from an exciting start into a quieter development but which were invigorated by advances in cognitive science in the 1990s. They now have a renaissance with the inclusion of human cognitive capabilities into the system to inform decision making. In addition, massive advances in the speed of computation in the decades since the definition of artificial intelligence gave a boost to cognitive computing and make these advancements possible.

The field of architecture provides excellent applications for artificial intelligence and cognitive computing. Architectural applications of artificial intelligence methods and techniques were introduced into design education as early as the 1970s and 1980s in the USA and later in Europe. Architecture requires the combination of structured input which is produced with rule-based systems, past experiences, and expectations of the future.

Urban design is an even more interesting application area of cognitive computing, since decision information derived from citizen input, such as transportation needs and external requests, is much higher than in individual buildings. This requires a better collection infrastructure for mining all opinions, proposals, and requests that can then be represented as data. This mix of requirements is almost identical with the computational tools and data available at present: structured input and requirements defined by city authorities; historical data and information needed as the basis for future design decisions; and varied user requirements. Together, these specifications help to define the future city or to improve the existing city.

Cognitive design computing is the combination of best practice from architectural design and urban design. From architecture, it takes efficient abstraction methods and knowledge about materials, climates, and the human use of the habitat going back several thousand years. From urban design, it takes the necessity to provide for large numbers of people that do not necessarily live in the urban system, but which rely on its infrastructure. Essential in both cases is that with the advent of big data and the improved capacity for analysis to observe and predict patterns and individual preferences, the urban design computing system will become more and more powerful.

As a logical next step, citizen design science will arise as a concept that adds the strength of thousands of citizens in terms of observation, human cognition, experience, and local knowledge into a scientific framework. Citizen design science can be described as the combination of citizen science and of design science. With citizen science, citizens of all ages and backgrounds support scientists by either collecting or analyzing data and observations. Millions of individual observations can then become usable data to improve the planning and functionalities of a city.

Companies, city governments, and universities discuss global cities, smart cities, intelligent cities, sustainable cities, or resilient cities. Something that might connect them all is the combination of the human cognitive system with advanced information technology, sensor information, and big data. Those, in turn, might develop into hybrid systems that might be best described as cognitive cities. The authors of this book are at the forefront of this new development that will shape the cities of the future.

Zürich   Gerhard Schmitt
February 2016

Contents

Editors and Contributors

About the Editors

Edy Portmann is a researcher, specialist, and consultant for semantic search, social media, and soft computing. At present, he is working as a Swiss Post-funded Assistant Professor of information science at the University of Bern. In the past, Edy Portmann studied for a B.Sc. in Information Systems at the Lucerne University of Applied Sciences and Arts, for an M.Sc. in business and economics at the University of Basel, and for a Ph.D. in computer sciences at the University of Fribourg. He was a Visiting Research Scholar at National University of Singapore (NUS), as well as a postdoctoral researcher at University of California at Berkeley, USA. During his studies, Edy Portmann worked several years in a number of organizations in study-related disciplines. Thus, among others, he worked as supervisor at Link Market Research Institute, as contract manager for Swisscom Mobile, as business analyst for PwC, as IT auditor at Ernst and Young, and, in addition to his doctoral studies, as researcher at the Lucerne University of Applied Science and Arts. Edy Portmann is repeated nominee for Marquis Who's Who, a selected member of the Heidelberg Laureate Forum, cofounder of Mediamatics, and coeditor of the Springer Series "Fuzzy Management Methods," as well as author of two popular books in his field. He lives happily married in Bern and has three lively kids.

Matthias Finger is known for his expertise in matters of regulation and governance of network industries. He holds a Ph.D. in political science and a Ph.D. in adult education from the University of Geneva and has been an assistant professor at Syracuse University (New York), an associate professor at Columbia University (New York), and a full professor of management of public enterprises at the Swiss Federal Institute of Public Administration. Since 2002, he holds the Swiss Post Chair in Management of Network Industries at Ecole Polytechnique Fédérale in Lausanne (EPFL), Switzerland. Since 2010, he also directs the Florence School of Regulation's Transport Area at the European University Institute in Florence, Italy. Since 2014, he directs the Institute of Technology and Public Policy at EPFL.

Prof. Finger is the coeditor in chief of the Journal *Competition and Regulation in Network Industries*, a member of the editorial board of *Utilities Policy*, and a member of the Swiss electricity regulatory authority (ElCom). His most recent global project focuses on the governance of large urban systems in collaboration with six global cities and partner universities, along with selected global firms (see: www.iglus.org).

Contributors

Ann Cavoukian is recognized as one of the world's leading privacy experts. She is presently the Executive Director of Ryerson Universities' Privacy and Big Data Institute. Appointed as the Information and Privacy Commissioner of Ontario, Canada, in 1997, Dr. Cavoukian served an unprecedented three terms as commissioner. There, she created Privacy by Design, a framework that seeks to proactively embed privacy into the design of technology, infrastructure, and business practices, thereby achieving the strongest protection possible. In October 2010, a Conference of International Privacy Regulators unanimously passed a resolution recognizing Privacy by Design as an essential component of privacy protection. Since then, PbD has been translated into 37 languages. Dr. Cavoukian's expertise has been recognized in many ways. She was ranked among the top 25 Women of Influence, named one of the top 100 City Innovators Worldwide by UBM Future Cities; chosen as one of the "Power 50" by Canadian Business magazine; and awarded an Honorary Doctor of Laws from the University of Guelph; she was selected for Maclean's Magazine's "Power List" of the top 50 Canadians and, most recently, picked as one of the top 10 women in data security, compliance, and privacy you should follow on Twitter. In her leadership of the Privacy and Big Data Institute at Ryerson University, Dr. Cavoukian is dedicated to demonstrating that privacy can and must be included, along with other considerations such as security and business interests. Her mantra of "banish zero-sum" enables multiple interests to be served simultaneously.

Michelle Chibba, in addition to providing privacy expertise for which her specialization is Privacy by Design, is also a research associate with the Privacy and Big Data Institute at Ryerson University (Toronto, Ontario). For the last 10 years, Ms. Chibba was airector, Policy Department and Special Projects at the Office of the Information and Privacy Commissioner of Ontario, Canada (IPC). During her tenure at the IPC, her department was responsible for conducting research and analysis, as well as liaising with a wide range of stakeholders to support Dr. Ann Cavoukian's leadership role in proactively addressing privacy and technology issues affecting the public, otherwise known as Privacy by Design. Michelle has over two decades of experience in strategy development, most of it in the public sector where she was responsible for developing and implementing several strategic policy projects. She received a master's degree from Georgetown University (Washington, D.C.), with a focus on ethics and international business. She is a frequent speaker on Privacy by

Design and emerging data privacy/technology issues and has written a number of publications on privacy and technology.

Sara D'Onofrio has got her bilingual Bachelor's degree in business administration from the University of Fribourg and her master's degree in business administration with specialization in information systems from the University of Bern. During her Bachelor's, she worked as an administrative assistant for a Swiss store chain for household appliances, consumer electronics, and computers. Before starting her Master's, Sara D'Onofrio worked for more than a year for a company that has been efficiently working on computerizing the Swiss Trade Registers. During her Master's, she worked as an administrative assistant for a Swiss painting company and later on as a research assistant for the Institute of Information Systems at the University of Bern, where she supported teaching programs and contributed to research on cognitive cities, cognitive computing, soft computing, and stakeholder management. Sara D'Onofrio has started her Ph.D. in business administration at the University of Bern in December 2015. Her research interests lie in the area of cognitive cities, cognitive computing, innovation management, requirements engineering, and soft computing.

Dominique Gadient has got two Bachelor's degree in business administration, one from the University of Bern and one from the University of Applied Science in Zurich, and a master's degree in business administration with specialization in Information Systems from the University of Bern. During his studies, he gathered experience in agile software development methodology (Scrum), and a deeper understanding about family businesses and entrepreneurship. After finishing his Master, he started a career as a consultant in an international management consulting company with focus on information technology topics. His research interests lie in the area of big data, digitalization, process mining, and soft computing.

Patrick Kaltenrieder is a research assistant/Ph.D. student at the University of Bern. After his Bachelor's degree and master's degree in business administration at the University of Bern, Patrick Kaltenrieder worked for about two years at a major Swiss Bank and then returned to the university for his dissertation. His research interests and fields focus on smart and cognitive cities; knowledge aggregation, representation, and reasoning; and stakeholder management and cognitive computing. Patrick Kaltenrieder is writing a blog about his smart and cognitive city research to keep the readers of his papers in the loop. The blog is available at http://smartandcognitivecities.blogspot.ch/. Patrick Kaltenrieder is the author of several papers published at various international conferences such as the FUZZ-IEEE 2015 in Istanbul.

Nalan Ilyda Karaca received the B.E. degree in industrial engineering from Kadir Has University, Istanbul, Turkey, and upon graduation conducted research for the university for two years. She pursued her studies at Stevens Institute of Technology, where she received her M.E. degree in engineering management. She received the Best Master's Thesis Award in smart city concept with "(2015) Smart City

Hoboken Energy Consumption Behavior of Citizens," under the supervision of her advisor, Dr. Mo Mansouri. She also volunteered as a research assistant at Stevens Institute of Technology. She has been working in the Information Technology Department of Credit Suisse (USA) LLC as an analyst, since July 2015.

Aigul Kaskina is a Ph.D. student at the Information System Research Group, University of Fribourg. In 2010, she finished an M.Sc. in computer science at the Oxford Brookes University, Oxford, UK. In 2009, she obtained a B.Sc. in computer science from Aktobe Regional State University, Aktobe, Kazakhstan. Her research focuses on voting advice applications, recommender systems, and fuzzy logic. She is a teaching assistant for the database course (Bachelor) and recommender systems course (Masters).

Mo Mansouri is an associate professor in the School of Systems and Enterprises and the program lead for Systems Engineering as well as Socio-technical Systems programs at Stevens Institute of Technology. He received his Ph.D. from The George Washington University in engineering management and his M.Sc. and B.Sc. in industrial engineering from University of Tehran and Sharif University of Technology, respectively. Prior to joining Stevens, he served several international development organizations and non-profits as a research fellow, program director, and consultant, working on strategic philanthropy and social entrepreneurship for development programs. He is published in many scientific journals and on a range of domains from smart solutions, and transportation, to financial and energy systems, among which are IEEE Systems Journal, Int. J. of Industrial and Systems Engineering, Journal of Operational Risk, Marine Policy, Maritime Policy and Management, Int. J. Ocean Systems Management, Enterprise Information Systems, Transportation Research Record, and numerous conferences particularly IEEE Systems Conference. His current research focuses on developing governance frameworks for effective policymaking and embedding resilience in complex networks and infrastructure systems.

Andreas Meier is a member of the Faculty of Economics and Social Sciences and a professor of information technology at the University of Fribourg. He specializes in electronic business, electronic government, and information management. He is member of GI (Gesellschaft für Informatik), IEEE Computer Society, and ACM. After studying music in Vienna, he graduated with a degree in mathematics at the Federal Institute of Technology (ETH) in Zurich, studied his doctorate, and qualified as a university lecturer at the Institute of Computer Science. He was a systems engineer at the IBM research laboratory in San Jose, California, director of an international bank, and a member of the executive board of an insurance company.

Robert Moyser is a director with BuroHappold specializing in large-scale master planning and urban regeneration projects both in the UK and globally. He also leads BuroHappld's smart city team specializing in strategy, procurement, and delivery. He is a key contributor within the industry on the topic of smart cities, and he is currently a member of two steering groups formed by the British Standards Institute to develop a new Publicly Available Specification (PAS 181) titled "Smart city framework – Guidance for decision makers in smart cities and smart communities" and a new Published Document (PD 8101) titled "Smart Cities: Guide to planning for new development." PAS 181 was recently completed and is the world's first guidance document on smart cities. He has also presented key note lectures at a number of smart city conferences both in the UK and abroad. He has a background in management consultancy and previously worked for Accenture. This broad experience was supplemented through the achievement of an MBA from Manchester Business School. Robert undertook this part-time from 2007 to 2011.

Marc Osswald is a Reporting and Analytics Expert at Swiss International Air Lines Ltd. Since his Bachelor's degree in information systems and his master's degree in information management at the University of Fribourg, he has been working in the airline industry for over four years. He is particularly interested in research about knowledge representation, cognitive computing, content verbalization, and fuzziness.

Elpiniki Papageorgiou is an associate professor in the Department of Computer Engineering at Technical Educational Institute, University of Applied Sciences of Central Greece, Lamia, Hellas. She has been working for over fourteen years as researcher in several research projects related with the development of novel computational intelligence methodologies for decision support systems, intelligent algorithms for decision making, data analysis and mining, and knowledge-based and expert systems. Recently, she has worked in four European Research Programs FP6 and FP7 (FP7-ICT-2013-11 ISS-EWATUS, FP7-ICT-2011 USEFIL, FP7-ICT-2007-1 DEBUGIT, FP6-SESAME), focused on the development of decision support systems using intelligent/cognitive methodologies, fuzzy cognitive maps, and data mining algorithms. She was the main organizer of the accomplished Special Sessions on Fuzzy Cognitive maps entitled as *Methods and Applications of Fuzzy Cognitive Maps* in previous FUZZ-IEEE conferences at 2011, 2012, 2013, 2014, and 2015 and has chaired four of them. She was also chair and main organizer of the 1st workshop on FCMs held at Paphos, Cyprus, in AIAI 2013. She is author and coauthor of more than 148 publications in journals, conference papers, and book chapters and has more than 1480 citations from independent researchers (h-index = 19 in scopus). Dr. Elpiniki Papageorgiou is the editor of the Springer book *Fuzzy Cognitive Maps for Applied Sciences and Engineering - from fundamentals to extensions and learning algorithms,* Intelligent Systems Reference Library 54, Springer 2014. Her research interests include expert systems, fuzzy cognitive maps, soft computing methods, decision support systems, computational intelligent algorithms, and machine learning.

Valerio Perticone is an Italian computer scientist. His interests are in social networks and fuzzy logic. He has authored several publications in international and national journals and proceedings. He is an Italian Wikipedia administrator and libre software advocate.

Marco Elio Tabacchi is a computational intelligence cognitivist, with interests ranging from the role of computational intelligence tools in development and deployment of smart cities to history and philosophy of soft computing and from complexity evaluation to recursive aspects of reality. He is currently the scientific director of Istituto Nazionale di Ricerche Demopolis and the director of operations at SCo2 (the research group for Soft Computing applications to Cognitive Science) at Università degli Studi di Palermo, Italy. He is serving as a member of the board at AISC, the Italian association for Cognitive Science. He holds a Ph.D. in physics and has authored more than 90 publications.

Luis Terán is currently working as assistant doctor and lecturer at the Information Systems Research Group, University of Fribourg, Switzerland, and he is a professor at Universidad de Las Fuerzas Armadas (ESPE), Ecuador. He earned a Ph.D. in computer science at the University of Fribourg (Information Systems Research Group). In 2009, he finished an M.Sc. in communication systems from the Federal Institute of Technology (EPFL), Lausanne, Switzerland. In 2004, he received a B.Sc. in electronics and telecommunications from Escuela Politecnica Nacional, Quito, Ecuador. His research interests include e-government, e-participation, e-collaboration, e-democracy, e-election, e-voting, e-communities, e-passports, recommender systems, and fuzzy classification.

Sabina Uffer is Head of Research Cities at BuroHappold undertaking research on urban policy and infrastructure planning. Before joining BuroHappold in 2013, she worked at LSE Cities on the project Resilient Urban Form and Governance conducting comparative research on urban neighborhoods in Hong Kong, Singapore, New York, Paris, London, and Berlin. Her doctoral dissertation investigated the process of financialization of Berlin's housing provision—from the political decision to privatize state-owned housing developments to the entrance of institutional actors through real estate private equity funds. She has taught at LSE and recently served as traveling faculty on IHP Cities in the twenty-first century: people, planning, and politics.

Marcel Wehrle is currently obtaining a Ph.D. in information and communication technology from the University of Fribourg, Switzerland. He received a M.A in communication and economic science and a premaster in informatics from the University of Fribourg, Switzerland. He is founder and co-owner of duckstance GmbH and emonitor GmbH and worked as IT project manager for 6 years. His research is focusing on granular computing, information granulation, self organizing maps.

Ronald R. Yager is professor of information systems and director of the Machine Intelligence Institute at Iona College. He is among the world's most highly cited researchers with over 47,000 citations to his work in Google Scholar. He is editor and chief of the International Journal of Intelligent Systems and serves on the editorial board of numerous journals. He has published over 500 papers and edited over 30 books in areas related to fuzzy sets, computational intelligence, human behavioral modeling, decision making under uncertainty, and the fusion of information. He is the 2016 recipient of highly prestigious IEEE Frank Rosenblatt award. He was the recipient of the IEEE Computational Intelligence Society Pioneer award in Fuzzy Systems. He received the special honorary medal of the 50th Anniversary of the Polish Academy of Sciences. He received the Lifetime Outstanding Achievement Award from International the Fuzzy Systems Association. He received honorary doctorate degrees, honoris causa, from the Azerbaijan Technical University and the State University of Information Technologies, Sofia Bulgaria. Dr. Yager is a fellow of the IEEE, the New York Academy of Sciences and the Fuzzy Systems Association. He has served at the National Science Foundation as program director in the Information Sciences program. He was a NASA/Stanford visiting fellow and a research associate at the University of California, Berkeley. He has been a lecturer at NATO Advanced Study Institutes. He is a visiting distinguished scientist at King Saud University, Riyadh Saudi Arabia. He is an adjunct professor at Aalborg University in Denmark. He received his undergraduate degree from the City College of New York and his Ph.D. from the Polytechnic Institute of New York University.

Noémie Zurlinden has got her Bachelor's degree in economics, social sciences, and philosophy and her master's degree in economics from the University of Bern. During her Master, she worked as a research assistant for the Institute of Information Systems at the University of Bern, where she contributed to research on soft computing. In the year 2015, she spent three months in Nairobi, Kenya, working at a research institute for Behavioral and Experimental Economics. She has started her Ph.D. in economics at the University of St. Gallen in August 2015. Her research interests lie in the area of development economics.

What Are Cognitive Cities?

Matthias Finger and Edy Portmann

Abstract "Smart city" as a concept is an appropriate and valuable answer to the efficiency challenges modern cities are facing today. Its epistemic foundations, however, rooted as they are in (command and) control theory and scientific management, lead to a very traditional and mostly technocratic view of urban management and government. Yet, the new urban challenges cannot be addressed solely by ways of increased efficiency. These challenges also—and probably mostly so—pertain to sustainability and resilience, requiring new and innovative approaches to urban governance. Such approaches will have to involve the "human factor", cognition, creativity along with the ability to learn so as to be able to deal with disruptive changes (resilience). In addition, cities are complex sociotechnical systems and it is therefore not possible to address their challenges thanks to technological developments and innovations only. In this chapter we will introduce a novel approach to overcoming the limitations of the concept of "smart cities" and explain the conceptual framework that underlies our approach, as well as the different chapters of this book. As such, we offer a broad and comprehensive perspective on so-called "cognitive cities."

1 Introduction

In this book we will introduce our readers to the concept of "cognitive cities." This is indeed a new concept, which has been used, as far as we know, by three sources:

M. Finger (✉)
Institute of Technology and Public Policy (ITPP), École Polytechnique Fédérale de Lausanne (EPFL), Route Cantonale, Station 5, 1015 Lausanne, Switzerland
e-mail: matthias.finger@epfl.ch

E. Portmann
Institute of Information Systems (IWI), University of Bern, Engehaldenstrasse 8, 3008 Bern, Switzerland
e-mail: edy.portmann@iwi.unibe.ch

© Springer International Publishing Switzerland 2016
E. Portmann and M. Finger (eds.), *Towards Cognitive Cities*,
Studies in Systems, Decision and Control 63, DOI 10.1007/978-3-319-33798-2_1

- In February 2011 a commercial Cognitive Cities Conference was held in Berlin (or at least announced, as we cannot tell whether it was effectively held). Its perspective was very broad and seems to have been inspired by urban planners and urban sociologists. The concept was not yet clearly defined at that time.
- In September 2011, Mostashari et al. published a 4-page paper in the Network Industries Quarterly (2011, vol. 33, no. 3, pp. 4–7) entitled cognitive cities and intelligent urban governance. They were clearly inspired by systems theory and placed the concept of cognitive cities within the context of the governance of systems, more precisely socio-technical systems.
- The technology consultancy firm IBM has started to use the concept of cognitive cities as of 2013. However, they seem to consider cognitive cities basically as a synonym for the concept of smart city, which we will discuss below. As of recently, IBM seems to favor the concept of "smarter cities".

The article by Mostashari et al. [8] seems to us to be, so far, the only intellectually solid conceptualization of cognitive cities and we will build on it here. In this introduction, we will first briefly present the concept of "smart cities", as this concept has emerged, so far, as the dominant concept in matters of city management and governance. To recall, many concepts have been tried on various occasions, such as "digital city," "intelligent city," "ubiquitous city," "creative city," "knowledge city," "learning city," and certainly others more. Yet, smart city seems to have imposed itself as the generally accepted concept in the literature at the present day. In a first section, we will therefore present this concept of smart city and discuss its limitations. In a second section we will then argue why cities need to become cognitive and therefore, why the concept of cognitive city is needed. Finally, we will highlight the main building blocks of cognitive cities. Overall, this introduction serves as the intellectual foundation of the concept of cognitive cities, a concept, which will be illustrated, in this book, by various cases.

2 Smart(er) Cities

Smart city has become the buzzword, mainly among (smart) equipment vendors, such as smart metering companies, telecommunications operators, location services, etc. IBM is without doubt the firm that has most pushed the concept of smart city, and more recently smarter cities, and has clearly the most precise definition of it.

IBM defines a smart(er) city as one that makes optimal use of all interconnected information available today to better understand and control its operations and optimize the use of limited resources [4].

In other words, smart cities build on the availability of data resulting from the more or less systematic rollout of the information and communication technologies (ICTs) in the various infrastructures and uses the so gathered information to better control, to better manage, to optimize ultimately to make infrastructure systems more efficient. This is of course a very valuable objective and, without doubt, can these infrastructure systems become "smarter" (i.e., made more efficient thanks to a systematic analysis of ubiquitous data and subsequent improved control over these systems).

IBM—and everyone else since—distinguishes five such urban (infrastructure systems), namely water, public safety, traffic, buildings and energy, leading, thanks to this approach, to "smarter water," "smarter public safety," "smarter traffic," "smarter buildings," and "smart energy." Of course, these systems are linked together and, at some point, must be integrated in order for urban systems to be optimally controlled and managed, thus the concept of "smart(er) city".

Let us state that this "smart city approach"—building on the systematic integration of data gather, bottom-up, from data points and sensors in the various infrastructures—is coherent, logical and indeed an appropriate answer to the efficiency challenge todays cities face. Indeed, resources—space, money, resources, skills, energy and time—are limited and the ICTs, combined with rational analysis and management tools (i.e., conventional hard computing tools and systems), can certainly squeeze more efficiency out of any of the above urban systems, as well as of the combined urban system as a whole.

But let us also state that the epistemological approach behind the smart cities concept is very traditional, rooted—as it is—in (command and) control theory and scientific management (i.e., top-down management). It is ultimately a purely engineering and thus technocratic approach to urban management and urban government (that is). While we do not dispute the internal coherence, the logic and the effectiveness of such an approach to addressing the urban efficiency challenge, we think that this approach has its limits and must ultimately be overcome—namely by the concept of "cognitive cities."

3 Why Do Cities Need to Become Cognitive?

In our opinion, there are three main reasons why the above (technocratic) approach to making cities smarter—while being appropriate for increasing efficiency—is not sufficient:

1. The first and main reason is that urban problems cannot be reduced only to problems of efficiency: while efficiency of urban infrastructure systems (incl. the integrated efficiency of the combined urban system) is a major challenge and worthwhile to be addressed, urban challenges also pertain to sustainability and resilience. Sustainability and resilience imply more and other approaches than the above technocratic one. Rather, they imply the inclusion of humans and ecology into the equation (sustainability), as they imply learning, creativity and disruptive changes (resilience). In other words, addressing urban challenges beyond efficiency requires learning and cognition and not just optimization.
2. The second, and probably equally important reason, why the smart city concept is insufficient, pertains to human beings and actors more generally. Indeed,

urban systems are not merely technical artifacts; rather they are complex socio-technical systems, whereby technology and institutions and organizations co-evolve. Even efficiency gains can only be achieved thanks to a combined technological and institutional/organizational approach, thus involving both the ICTs and (social) actors. It is obvious that cognition is ultimately located within actors, as well as in the interaction between the actors themselves and between the actors and the technological systems.

3. Thirdly the smart city approach reduces the use of the ICTs to a purely technocratic and optimizing tool, thus underestimating the potential of the ICTs. Indeed, the ICTs have already proven to be useful for cognitive purposes (and not just for optimizing purposes), as evidenced collective intelligence and computation intelligence approaches (i.e., nature-inspired computational methodologies [6]). In other words, the ICTs can (and must) also be used for collective learning, especially learning of and by urban systems, rather than simply for optimizing urban infrastructure systems (i.e., making them smarter [7]).

For all above three reasons combined, we consider that cognitive cities is not only a necessary but—as an enhancement—also a more appropriate concept, especially when it comes to addressing the challenge of urban resilience. Let us explain.

4   What Are the Cognitive Cities' Various Components/Building-Bocks?

To be clear: cognitive cities build on learning cities, which in turn build on smart cities; they do not replace each other. Our approach and argument is structured into three layers, along the three main challenges that we think cities are facing today (i.e., the efficiency, the sustainability and the resilience challenges):

- Cities are undeniably facing efficiency challenges: they have limited resources with which to satisfying exponentially growing (energy, transportation, water, etc.) demands. Duplicating the physical (energy, transportation, water, etc.) infrastructure layers by way of an information or data layer is a first and necessary step to address this efficiency challenge. Of course, this data or information layer is itself only possible because there is an underlying (tele) communications infrastructure layer rolled out in cites (e.g., fiber, Internet of things (sensors), mobile antennas, switches, etc.). Analyzing the so generated data and information in a structured and systematic fashion from a business process optimization (operations research) perspective makes perfect sense and helps to optimize these infrastructure systems, individually and as a whole integrated urban system, so as to make cities for efficient, thus "smarter" (smart city).

- But cities also face <u>sustainability</u> challenges: beyond become more efficient, they also must become economically, socially and ecologically more sustainable. Mainly because of the so-called "rebound effect," this cannot be achieved by efficiency measures alone and involves substantial behavioral change of all involved actors, both individuals and (non-profit) organizations. They must reduce, they must be actively involved in changing their work, social relations and consumption patterns. This is a qualitative and not just a quantitative change, even though it is gradual and not disruptive. The data and information generated thanks to the ICTs (see above) can and must be made available to the different actors (individuals and organizations) so that they can be used by them for learning and ultimately behavioral change purposes. Tools for data analysis (e.g., big data and/or business analytics) and display, along with social media tools, contribute to the actors' individual and collective learning, something we may call "<u>learning city</u>".

- But facing the *resilience* challenges goes still one step further and requires yet another level of cognition: indeed, in our complex world, urban systems—technology and actors combined—must ultimately become more resilient, that is capable of withstanding external shocks (i.e., economic crises, epidemics, heat waves, water shortages, congestions and transport breakdowns, specific environmental pollutions, riots, etc.), something which goes much further than increasing efficiency or sustainability. Building on the above two layers—data generation and analysis for optimization purposes, as well as data display and exchange for individual and collective learning purposes [7, 11]—new forms of machine intelligence involving human-machine interactions will be used to come with creative and disruptive systemic solutions, by which (the) entire urban socio-technical system(s) copes or adapts to shocks from its environment. The basis of these systems will allow cities and its citizens to evolve. This is what we call "<u>cognitive city</u>," and for the realization of which we explore, in this book, the appropriate (socio-technical) tools. More precisely, for cities to compete effectively in today's knowledge-based society, citizens need to align practices with the changed characteristics and context of knowledge. Hierarchical, top-down, pre-defined methods of operating fail to react and adapt quickly to changes [11]. Any variation of the foundations of complex large urban systems (i.e., by exposure of the city to external shocks) can shift overnight—rendering today's knowledge predominantly obsolete. Therefore, a new thinking inspired by complex dynamic systems, ecosystems, and social systems should be welcomed. Yet, such a perspective would also have implications for the robustness of an urban system. More and more, for cities to survive, they must increasingly rely on learning networks and knowledge ecologies [2].

In recent times, that is to say, learning started to change more and more. Indeed, conventional theories (as behaviorism, cognitivism and constructivism) offer different ways of looking at learning and cognition, however, they most often fail short with informal, networked and technological learning [1]. They leave open what adjustments need to be made when technology performs many of the cognitive operations previously performed by human beings (e.g., data, information, and knowledge storage as well as retrieval). By contrast, connected learning theories [1, 5]—and the incidental connectivism [11]—take experience and emotion into account for sense-making [14]. Among other things, connectivism that way allows a cognitive city (as well as its citizens) to become more resilient (i.e., to "bounce back" from shocks).

At that, the starting point is always the individual citizen. Personal knowledge is comprised of a network, which feeds into cities organizations and institutions, which in turn feed back into the network, and then continue to provide learning to individual citizens. This cycle of knowledge development (personal to network to organization/institution) allows learners (i.e., individual citizens as well as institutions/organizations) to remain current in their field through the connections they have formed [10]. With this in mind, ICTs help evermore to connect individual citizens (e.g., with each other but also with organizations/institutions). Connectivism addresses the challenges that cities face in knowledge management activities (e.g., embodied data and information that is embedded in representations of interaction [9], coordination of enaction among embodied agents [12], and ecological contributions to a cognitive ecosystem, etc.). The application thereof leads to some kind of a city's (distributed) cognition (with knowledge not only lying within the individual, but also in it's social and physical environment). Thereby knowledge that also resides in smart computing tools and systems (i.e., database or (Web) knowledge bases) needs to be connected with the right people in the right context. To contrive this is where cognitive computing systems come in.

5 What Are Cognitive Computing Systems?

According to Hurwitz et al. [3], a system to become cognitive comprises three fundamental principles:

- A cognitive system learns: To this end, the system leverages data to make inferences about a domain, a topic, a person, or an issue based on training and observations from all varieties, volumes, and velocity of big data.
- To learn, the system needs to create a model or representation of a domain (which includes internal and potentially external data) and assumptions that dictate what learning algorithms are used: Thereby, understanding the context of how the data fits into the model is key to a cognitive system.
- A cognitive system assumes that there is not a single correct answer: The most appropriate answer is based on the data itself. A cognitive system uses these data to train, test, or score a hypothesis.

Now, when data is acquired (e.g., over traditional ICT systems), curated and analyzed, the cognitive system must identify and remember patterns and associations in the data. This iterative process enables the system to learn and deepen its scope so that understanding of the data improves over time. One of the most important practical characteristics of a cognitive system is the capability to provide the knowledge seeker (or rather the citizens permanently interacting with the system) with a series of alternative answers along with explanation of the rationale or evidence supporting each answer. This initiates a mutual communication process with the system (with its strengths in big data and analytics), which amplifies humans (with its strengths in emotion).

To team up with human beings in a (more) natural and seamless way (e.g., rooted in biological processes; which show by their very existence their ability to cope with complexity [6]), thus a cognitive computing system consist of tools and techniques, including machine learning, Internet of things, natural language processing, causal induction, probabilistic reasoning, and data visualization (cf. the various cases presented in this book). The term cognition thereby bespeaks already the systems role model—the human brain. Such use of role model is also referred to as soft computing [15], a toolbox of methods and techniques which allows a step from (hard) technical systems towards (soft) human beings.

6 Addressing the Cities' Efficiency, Sustainability and Resilience Challenges

Taking advantage of a cognitive system's synergy with (social) human beings in a city, soft computing enhances traditional (hard) computing techniques (i.e., most often used by today's smart city vendors) by a tolerance towards imprecision, uncertainty, partial truth, and approximation. Exploiting its constituent's methods (e.g., fuzzy logic, neural networks, evolutionary computing, etc.) as an enhancement of traditional systems able to naturally (e.g., following a biomimetic method [6]) connect human and computer system by ICT, it is predestinated to address the identified city challenges (i.e., efficiency, sustainability and resilience):

- Already traditional hard computing techniques allow for building smart cities, which are addressed today (e.g., by IBM Smarter City Initiative). On that score, many of today's cities concentrate on developing processes to become more efficient. Thereby, these cities are (too? [13]) often assisted and advised by (smart) equipment vendors (as already pointed out).
- Using a suitable learning theory (as connectivism [11]), the interaction of cognitive (soft computing) systems with citizens allows for building learning cities. As individual human actors can learn based on these systems (e.g., trough on-going human-computer-interaction), this supports citizens acting more sustainable (i.e., change bad behavior for good one in respect to ecology, economy, and society). Moreover, since soft computing focuses on systems able to process uncertain, imprecise and incomplete data and information—thus achieving adaptive, robust, and low solution cost as not everything has to be calculated to the final decimal—allows for more sustainable cognitive systems.

- The mélange of—by connected learning aligned—cognitive actors (i.e., citizens, cognitive systems, government bodies, institutions, organizations, politics, etc.) leads to <u>cognitive cities</u>, as an emerging kind of superorganism (i.e., a city consisting of as many different actors as possible, which follow somehow biological properties as emergence, learning, evolution, networks, fitness, autonomy, tolerance, etc. [6]). Thereby every single actor (incl. the cognitive system(s) itself) in this city is empowered by ICT networks to develop autonomously, which allows building the city's resilience (e.g., ecological, organizational, psychological, etc. ability to cope with change). The superorganism' underlying notion affords the city (very similar to what human beings are doing it) to learn (i.e., to connect fields, ideas and concepts). All these connections may be viewed as a kind of network, as a new connection is nothing more than a new-formed combination of—via dendrites and axons connected—neurons that fire together. This forms a new configuration, which had never been formed in this way. Based on a continuous perception and action cycle, the brain cells (as well as the body) organize itself constantly new.

The ability of the brain to adapt continuously, which we take into account for and by the implementation of cognitive cities, is called neuroplasticity [1]. Hence, to move a first step from smart cities, it makes sense to follow a transdisciplinary research approach to get to the bottom of cognitive cities eventually. Our proposed approach thereby is twofold: For one thing we concentrate on action research, a process of inquiry transdisciplinary (i.e., beyond academic disciplines as well as beyond academia itself) conducted by and for those taking the action. Here, the primary reason for engaging in action research is to assist the actors (i.e., citizens, cognitive systems, government bodies, institutions, organizations, politics, etc.) in improving and/or refining his, her or its actions. For another thing we use (bio-mimetic [6]) design science research to shape these cognitive cities. This research involves the design of artifacts and the analysis of the use and/or performance of such artifacts to improve and understand the behavior of aspects of the city.

7 Outline of Our Book

This volume features nine chapters illustrating various aspects and dimensions of cognitive cities. The logic of its structure proceeds from more general considerations to more specific illustrations. Follows a brief description of each of the chapters contained in this volume.

Robert Moyser and Sabina Uffer's chapter entitled "*From Smart to Cognitive: A Roadmap for the Adoption of Technology in Cities*" builds on the academic literature on smart and more recently cognitive cities, as well as the authors' own experiences in implementing technology infrastructure, to explore the role of technology in helping to create successful cities. It analyzes the challenges that technological solutions carry and provides a roadmap for the adoption of technology. The chapter concludes by stating that technology is simply one part of city systems. To create truly livable cities, city administrators need to develop a vision, engage the community and stakeholders, and undertake their due diligence in adopting technology.

Luis Teran, Aigul Kaskina and Andreas Meier's chapter entitled "*Maturity model for cognitive cities: three case studies*" focuses on the interaction between administrations and citizens. It highlights an eGovernment framework through which eEmpowerment of citizens can be achieved via the promotion of their participation. With the help of the ICTs, smooth communication between governments and citizens can be established, thus facilitating collective decision-making processes. The chapter illustrates this by way of three case studies, namely a collaborative legislation process, social voting as an illustration of eDemocracy, as well as civic smart participation.

Ann Cavoukian and Michelle Chibba's chapter entitled "*Cognitive Cities, Big Data and Citizen Participation: the Essentials of Privacy and Security*" argues that privacy cannot be sacrificed for other anticipated benefits. Privacy protections will indeed be critical to the adoption of cognitive city sensor technologies. The authors state that individuals must feel comfortable that their privacy will not be violated as they move about in public spaces. Consequently, both privacy and functionality need to co-exist. In other words, adopting privacy early on already at the design phase will be necessary and "Privacy by Design" is the sine qua non for all advances in technology, data management and application of cognitive technologies, as envisioned by cognitive cities.

Valerio Perticone and Marco Elio Tabacchi's chapter entitled "Towards the improvement of Citizen Communication through Computational Intelligence" starts out from the observation that dealing with problems that arise from collective sharing of resources in metropolitan areas generally leads to interactions between citizens and local governments by way of natural languages. Digital technologies increasingly allow residents to better communicate with administrators, legislators, planners, etc. who routinely use technical terminology seldom accessible to the layperson, or linguistic styles that are not immediately understandable. The chapter demonstrates how computational intelligence approaches can be used in order to improve such communication and discusses particular applications that help reduce the communication gap between citizens and government.

Patrick Kaltenrieder, Elpiniki Papageorgiou and Edy Portmann's chapter entitled "*Digital Personal Assistant for Cognitive Cities: A Paper Prototype*" presents an evaluation and initial testing of a meta-application for enhanced communication and improved interaction between stakeholders citizens. The authors argue that this meta-app has the potential to improve the living standards of citizens, helping them to more effectively manage their time and organize their personal schedules.

Marc Osswald, Marcel Wehrle and Edy Portmann's chapter entitled "*Verbalisation of Dependencies Between Concepts Built Through Fuzzy Cognitive Maps*" outlines a method for enhancing communication between machines and humans trough verbalization of causalities in fuzzy cognitive maps (FCMs). The chapter furthermore introduces a verbalization technique that relies on the restriction centered theory (RCT) of reasoning which is able to handle the imprecision of language by introducing restrictions. As proof of concept, the authors' analyses taxi passenger flows and transform the causalities of the FCM into words with the help of the RCT.

Sara D'Onofrio, Noémie Zurlinden, Dominique Gadient and Edy Portmann's chapter entitled "*Cognitive Cities: An Application for Nairobi. Text Message Participation of Slum Inhabitants to Improve Sanitary Facilities*" illustrates, based on a use case, how a city in an emerging country can quickly progress using the concept of smart and cognitive cities. The chapter shows how slum inhabitants can inform the responsible center via text messages in cases when toilets are not functioning properly. Through cognitive computer systems, the responsible center can fix the problem in a quick and efficient way by sending repair workers to the area. Focusing on the slum of Kibera, an easy-to-handle approach for slum inhabitants is presented, which can make the city more efficient, sustainable and resilient.

Mo Mansouri and Nalan Ilyda Karaca's chapter entitled "*Innovative Urban Governance: A Game Oriented Approach to Influencing Energy Behavior*" applies theories of interactive collaboration to a simple yet effective game, which involves citizens of an isolated environment with dynamic adjustment of their behavior in regards to energy consumption. The case of Hoboken focuses on the evaluation of behavioral change among residents of Stevens Institute of Technology, given the right collective information and provided the sustainable incentive structure. As a part of this research, the results of the experiment with on campus residents were analyzed against similar information collected from citizens of Hoboken. The results of the research supports the hypothesis that people will choose a sustainable alternative when given the right information and provided with incentives to do so.

Ron Yagers chapter entitled "*Diversity Measures for Smart Cities*" presents a mathematical discourse about possible diversity measurements, an important element of future megacities fairness (e.g., in and for participation and distribution purpose). For this very reason, such criteria might, for one thing, be used for smart compositions of mixed groups or, for another, be included in an enhanced political decision-making (e.g., using the idea of linguistic values and fuzzy sets to do so). In cognitive cities, diversity and fairness will be an important governmental consideration, Ron addresses with his proposed measurements.

These nine chapters, taken together, do indeed offer a comprehensive view of the different research endeavours about cognitive cities and hopefully will help pave the way for this new and innovative approach to governing cities in the future.

References

1. Caine, R., Caine, G.: Natural learning for a connected world: education, technology, and the human brain. Teachers College Press (2013)
2. Helbing, D.: Thinking Ahead-Essays on Big Data, Digital Revolution, and Participatory Market Society, Springer (2015)

3. Hurwitz, J., Kaufman, M., Bowles, A.: Cognitive Computing and Big Data Analytics. John Wiley & Sons (2015)
4. IBM: Smarter cities series: a foundation for understanding IBM smarter cities. Redbooks: REDP-4733-00 (2011)
5. Ito, M., Gutiérrez, K., Livingstone, S., Penuel, B., Rhodes, J., Salen, K., Schor, J., Sefton-Green, J., Craig Watkins, S.: Connected Learning. Digital Media and Learning Research Hub (2013)
6. Kaufmann, M., Portmann, E.: Biomimetics in design-oriented information systems research. In: Donnellan, B., Gleasure, R., Helfert, M., Kenneally, J., Rothenberger, M., Chiarini Tremblay, M., Vandermeer, D., Winter, R. (eds.) At the Vanguard of Design Science: First Impressions and Early Findings from Ongoing Research Research-in-Progress Papers and Poster Presentations from the 10th International Conference, DESRIST. pp. 53–60. Dublin, Ireland 20–22 May 2015
7. Malone, T., Bernstein, M.: Handbook of Collective Intelligence, MIT Press (2015)
8. Mostashari, A., Arnold, F., Mansouri, M., Finger, M.: Cognitive cities and intelligent urban governance. http://newsletter.epfl.ch/mir/index.php?module=eples&func=getFile&d=240&inline=1 (2011)
9. Pfeifer, R, Bongard, J, Iwasawa, S.: How the Body Shapes the Way We Think: A New View of Intelligence, The Mit Press (2007)
10. Siemens, G., Connectivism: A Learning Theory for the Digital Age, International Journal of Instructional Technology and Distance Learning, Vol. 2 No. 1, Jan 2005
11. Siemens, G.: Knowing Knowledge. Lulu.com (2006)
12. Thompson E.: The Enactive Approach: Mind in Life: Biology, Phenomenology, and the Sciences Of Mind, Harvard University Press (2010)
13. Townsend, A.M.: Smart Cities: Big Data, Civic Hackers, and the Quest for a New Utopia. W. W. Norton & Co, New York (2013). ISBN 978-0-393-08287-6
14. Weick, K.: Sensemaking in Organizations, Sage Publications Inc (1995)
15. Zadeh, L.A.: Fuzzy logic, neural networks and soft computing. Commun. ACM **37**(3), 77–84 (1994)

From Smart to Cognitive: A Roadmap for the Adoption of Technology in Cities

Robert Moyser and Sabina Uffer

Abstract With ever more of the world's population living in urban areas, cities face immense challenges ranging from providing basic human needs, to adapting to the impact of climate change, to improving the quality of life for its citizens. Addressing these issues, cities increasingly turn towards technology as the solution. Advancement in information communication technologies allows the collection of data from urban infrastructure systems relating to people's consumption patterns, therefore enabling resource optimization and more informed decision making. The adoption of technology however also brings many challenges. Taking leads from the academic literature on smart and—more recently—cognitive cities as well as the authors' own experience in implementing technology infrastructure, the chapter explores the role of technology in helping to create successful cities, analyses the challenges that technological solutions carry, and provides a roadmap to the adoption of technology. The chapter concludes that technology is simply one part of city systems. To create truly livable cities, city administrators need to develop a vision, engage the community and stakeholders, and undertake their due diligence in adopting technology.

Keywords Smart city · Cognitive city · Information communication technology · Smart infrastructure · Living City

1 Introduction

The future will be more urbanized than ever. Since 2010, more than half of the world's population live in urban areas. The UN estimates that by 2050, this number will jump to over 70 % [21]. As urban migration continues, cities face immense

R. Moyser (✉) · S. Uffer
BuroHappold, London, UK
e-mail: robert.moyser@burohappold.com

S. Uffer
e-mail: sabina.uffer@burohappold.com

© Springer International Publishing Switzerland 2016
E. Portmann and M. Finger (eds.), *Towards Cognitive Cities*,
Studies in Systems, Decision and Control 63, DOI 10.1007/978-3-319-33798-2_2

challenges ranging from meeting basic human needs for clean water, food, and shelter to dealing with waste and pollution, to providing mobility and access to jobs and services, and meeting aspirations for a high quality of life and greater well-being. Climate change and its immediate and future impacts are aggravating these challenges. Cities account for around 70 % of global energy consumption and over 70 % of energy-related carbon emissions [7]. Similarly, waste generation is foremost an urban issue and is therefore accelerating in parallel with urbanization. According to a World Bank report, urban solid waste generation will increase 70 % in the next decade [24]. This not only affects our natural environment and is directly linked with natural resource depletion, but ultimately also impacts upon our health and well-being.

While cities and the people living in them generate these problems, they are increasingly also considered to be the solution to them (see, for example, [11, 21]). It is cheaper and more resource efficient to provide public transportation, housing, electricity, water, and sanitation on a per capita basis for a densely settled urban area than for a dispersed rural population [21]. Cities also have the economic and social capital needed for rapid and radical change.

City authorities increasingly understand that urban environments are complex, adaptive systems of systems in which economic, social, spatial, environmental, and infrastructural systems must be managed in an integrated manner. Meeting the requirement for enhanced outcomes in terms of quality of life on the one hand and greater resilience (successful adaptation to fast and slow moving shocks and stressors) on the other, necessitates ever greater sophistication of governance. To tackle these urban challenges, technological solutions are being increasingly promoted under buzzwords such as 'smart cities', 'digital cities', 'intelligent cities', or—the latest addition to the debate—'cognitive cities'.

Some [1, 12] argue that the digital revolution may transform our lives in the same way the industrial revolution has over the past two centuries. Information communication technologies enable ubiquitous connectivity and super-fast access to Internet at ever lower cost. Increasingly, the technology not only allows for communication between people, but also between sensor-embedded digital devices and databases [19]. This ubiquitous use of technology is changing the way we live, work, shop, and play. The greatest change has occurred at an enterprise level. In business, information technology has overhauled how companies and entire industries operate. The fact that financial transactions can now be undertaken in a matter of milliseconds has completely transformed the banking industry. Similarly, in buildings, advances in design through intelligent systems have improved resource efficiency. On an individual level, the rise of smartphones has revolutionized how we communicate and travel. Advances in human cognition and artificial intelligence have the potential to further transform our work and lives through extending machine-to-human interactions to machine-to-machine interactions.

While the advancement in technology has been scaled up to support cities (e.g., real-time information on public transport services to sensor embedded energy grids), there is still very limited demonstration of integrated information communication systems across city departments and between stakeholders [15]. There are a number of reasons for this and they cover political, regulatory, economic, social, and technological challenges. Cities and their departments are inherently complex systems; decisions take time, involving multiple stakeholders, and procurement processes that do not always align with the private sector and their timeframe to respond [19]. The integration of information communication technologies is thus not as simple as adopting a new vendor together with their proprietary technology. Moreover, cities will need to think beyond information communication technology to issues such as equality, sustainability, and quality of life.

Addressing the question of why cities have not fully embraced the opportunity offered by technology and what is needed to adopt technology, is the subject of this chapter. After a short introduction to the concepts of smart and cognitive cities as well as the role of information communication technology infrastructure, the chapter examines the barriers for the roll-out of technology across city systems. It then outlines the conditions required for a successful city—be it smart or cognitive. In doing so, the authors argue that, for smart or cognitive cities to be successful, technology alone is not sufficient but needs to be integrated in broader strategic planning and urban management practices. Finally, the authors propose a roadmap for the implementation of technology to achieve more livable, sustainable, and resilient urban environments.

2 From Smart to Cognitive

To understand the relatively new concept of the cognitive city, it makes most sense to start with a definition of the smart city. While there is no single accepted definition of the concept, smart cities (and their variants) can be defined as the leveraging of information communication technologies such as satellites, wired and wireless technology, sensors, and other forms of data collection devices to help citizens, businesses, and governments to make data-driven, intelligent decisions to address environmental, economic, and social issues [10], Townsend [20]. Smart city initiatives can support better public services and infrastructure, access to social and economic opportunities, personal safety and security, effective healthcare, efficient transport systems, and—maybe most importantly—foster more resource efficient behavior among citizens.

Among a small number of academics, smart city debates have been pushed further by introducing cognitive theory. Cognitive theory, in a nutshell, posits how knowledge is produced through memory creation based on experience and observation. The still limited literature relating to cognitive cities defines the concept as "an extension of smart cities using cognition theory" [9: 1] in the sense that implies the "existence of learning, memory creation and experience" embedded in a city such that technology is leveraged to continuously improve the way it works. Learning, in this case, is not only undertaken by humans alone, but by various information holders (e.g., smart phones, computer systems, infrastructure operation systems, etc.). A cognitive system is able to sense, perceive and respond to changes in the environment and can therefore improve a system's performance by increasing its adaptive capacity [13: 4]. This means information technology as well as human cognition is used to adapt and respond to changes in the environment in order to improve urban governance.

In more concrete terms, in the smart city, individual citizens predominantly receive information on urban infrastructure such as traffic conditions or service outages; in the cognitive city, they also deliver information to others (e.g., other devices and sensors, machines, operating platforms, humans) to allow these systems to learn from and adapt their behavior [13]. This allows citizens to move towards a more efficient and sustainable use of resources and it allows service providers such as utility companies or transportation authorities to continuously adapt to provide more efficient and cost-effective services [10]. One such example, at an individual level, is the fact that our devices (i.e., our smart phones, tablets, etc.) will increasingly be able to interact with each other. If someone's device notes that the person is late for an appointment due to traffic or a previous appointment, it will automatically alert the person's next meeting's participants and recalibrate time and location for the meeting [8].

The technological possibilities to enhance our lives are exciting and advances in technologies represent an obvious area to explore, embrace, and adopt. The growth in access to data and improved communications, for instance, can provide the opportunity for citizens to interact with each other more efficiently and to establish greater engagement and transparency between themselves as well as those that manage communities, towns, or cities on their behalf. Technology has the potential to make urban dwellers' lives easier, more convenient, and less wasteful. It allows city governments to make more informed choices based on real-time data gathered. It enables more efficient service provision through better demand and supply management of key resources such as energy and water. It can make provision of utilities more resilient with less disruption of services. In addition, beyond the planning and automation of infrastructure, digital technology has the enormous economic potential of allowing the creation of new industries centered on citizen centric services, providing critical communication vehicles for citizens and communities to support their daily lives. It can help bring communities together and support them both in their daily routines as well as in extreme situations such as disaster.

Leveraging technology however also carries with it risks and the potential of unintended consequences. The creation of successful cities—cognitive or not—thus requires more than just technology. It requires the careful interaction of human and artificial intelligence across different urban systems. This includes exercising caution over how and by whom technology is leveraged. Authors such as Townsend [20] or Greenfield [5] argue for more 'grassroots' approaches to the cognitive city where citizens are empowered to use technology to improve their quality of life. It should be remembered that, first and foremost, cities are places for people—and one can take from this the fundamental principle that everyone within a city should be able to take benefit from the adoption of technology.

2.1 Role of Information Communication Technology (ICT)

The advances made in the field of Information Communication Technology (ICT) are enabling the development of cognitive cities. ICT has become the fourth utility sitting alongside the other three utility infrastructures (electricity, gas, and water) which are crucial to the delivery of successful developments in existing and new cities. ICT provides the infrastructure that enables communication through national and international voice, video, and data interconnectivity. Below and above ground infrastructure communication technologies provide an urban 'nervous system' and an interface between the city's primary assets and its end users. This nervous system then acts as an integrator for other city infrastructure layers.

Typically, implementation of ICT systems is categorized according to the seven layers of the Open Systems Interconnection (OSI) model developed by the International Organization for Standardization (ISO). In short, there is a physical layer that refers to the deployment of fixed (e.g., fiber, ADSL) and wireless (e.g., WLAN) data communications and telecommunication outside plants (OSP). It incorporates the physical space allocated to cable routes, underground civil infrastructure, street furniture, cellular and wireless antennae and access points (AP's), central plant and equipment rooms. On top of this 'hardware' is a range of 'soft' layers such as network operating systems, data-conversion utilities, and applications that enable the successful exchange of data from infrastructure to end user. Whilst the hard and soft layers of ICT provide the infrastructure, other considerations such as governance, procurement, asset lifecycle planning, and the business case are equally important to deliver an appropriate ICT architecture for any adopting city. The design of efficient and resilient ICT systems is key for the success of cognitive cities.

2.2 Application of ICT Infrastructure

A well planned and future-ready physical ICT infrastructure system forms the backbone that allows for sensor installations, integration of applications, and ubiquitous data collection. This, in turn, permits the development of a wide range of city improvement initiatives. These can generally be distinguished as top-down or bottom-up initiatives. Top-down initiatives focus on the wide use of information communication technology integrated with city systems "that enable central planning and an integrated view of the processes that characterize urban operations" [16: 7]. Traditionally, these initiatives are implemented by partnerships between city government and technology companies. The focus lies on smart city initiatives that improve the efficiency of infrastructure systems and services such as energy, waste, water, environment, and transportation. IBM and the City of Portland, USA, for example, collaborated to develop a computer simulation to understand how the city's core systems work together and, in turn, identify smart city opportunities [6].

The top-down approaches of corporations has however led to a wide distrust by cities and citizens who have become increasingly skeptical of the broader opportunity, due to the often high cost, lack of transparency, appearance of closed market place, and little discernable benefit for citizens.

Bottom-up initiatives, in contrast, are decentralized by definition. Whilst still relying on ICT, they emphasize the ways in which citizens generate, collect, and make use of the data. Initiatives focus on the improvement of the user experience and the enhancement of quality of life. They are typically developed by service providers, often software companies and start-ups, which are targeting consumers or end users Townsend [20]. Examples include applications that provide citizens real-time information to indicate the fastest (and cheapest) route to travel at any given time or on-demand and instant services such as one-click food orders, car-, or laundry-services. Problems associated with bottom-up approaches are the lack of joined-up systems, the time it takes to make a broad and meaningful contribution to city life, and the rate of failure of such initiatives Townsend [20]. In both cases, the top-down and bottom up approaches, the hard and soft infrastructure of information communication technology plays a key role.

These initiatives—both bottom up and top down—have tended to remain singular, localized interventions in urban governance and development. Moreover, these initiatives do not necessarily go beyond the 'smart city' approach defined above. Cognitive city approaches and initiatives where the capacity of learning and experience is integrated into the software, application, or 'machine' is harder to come by and is only in its very early stages with applications such as Waze or Snips[1] [8]. The following section explores why neither smart city nor cognitive city initiatives have yet managed to take a greater role in urban development and governance. This is analyzed through the lenses of political, regulatory, economic, and social challenges.

3 Challenges to Adopting Infrastructure Technology for Smart or Cognitive Cities

Compared to buildings, businesses, or even entire industries, cities are far more complex with a wide range of social, economic, and environmental networks and systems interacting with each other on a daily basis. They therefore present a multitude of challenges to the implementation of city-wide technology infrastructure, be it for smart or cognitive systems. These challenges can be categorized as political, regulatory, economic, social, and technological.

---

[1]https://www.waze.com/; http://snips.net/ [accessed 02.10.15].

3.1 Political Challenges

City administrations face enormous pressure from citizens and businesses to deliver a higher quality of life and an enhanced business environment and to do so equitably, responsively, and transparently. At the same time, limited resources, both environmental and financial, constrain their options, demanding a high degree of effective governance capable of prioritizing and implementing technology infrastructure rapidly.

In meeting this need, cities face three key issues. First, urban sprawl in the twentieth century was not always accompanied by a comparable extension of the administrative boundaries of city governments. While citizens cross these administrative boundaries on a daily basis through commuting to work, attending cultural events, or through their social ties, city administrations do not often have the necessary authority to govern or implement projects beyond their own boundaries. For example, there may be individual public authorities, such as the Port Authority of New York and New Jersey, which coordinate the air and water transportation in the metropolitan region, but there is no overarching authority for governing the metropolitan area of New York [14]. The mismatch between the urbanized areas and jurisdictional boundaries often hinders procurement and implementation of new or upgrading of existing infrastructure to smarter or more cognitive systems.

The second political challenge cities face is that, over the last century, they have become increasingly organized along functional structures with sector-specific agencies and city departments providing individual public services. The water provider(s), the transportation agency(ies), the energy provider(s), or the housing department operate each within their own organizational siloes and with their own departmental management structure and budgets. This creates an isolated and highly channeled approach to service provision and complicates the introduction of new technology where high set-up costs may be involved. Pressures to reduce the size of governments with the consequence of municipal services increasingly being contracted out to the private sector and third-party organizations aggravate the issues, mainly because of a lack of mechanisms for the transfer of information [14]. An often cited example is that when streets get opened-up several times a year to provide access for the different utility providers, other relevant city departments do not coordinate their maintenance efforts. Even when advanced information communication technologies have been adopted, the embedded organizational structure often hinders learning and memory creation across sectors.

Another political challenge is the alignment of timescales and the actual time horizons for transformation. City leadership and planning processes have often a lifecycle of four to five years which restricts their ability to act at the speed of technological transformation which may be as short as a few months or one to two years. At the same time, city leaders are rarely thinking beyond the election cycles, which goes against the needs of long-term coordinated land use and infrastructure planning, which require a twenty year or even longer perspective. Processes of planning, delivery, and experiential learning are thus interrupted by political timelines.

3.2   Regulatory Challenges

The use of technology will not only increase the volume of data collected, but also the complexity of how systems interoperate with each other (e.g., traffic control system with energy control system—therefore creating a 'system of systems') [2]. This poses a range of regulatory challenges, predominantly in the domain of data privacy, data security, and commercial liability when things go wrong. Current regulatory frameworks and legal policies are often not sufficient enough to deal with ownership of data, privacy protection, and security breaches. This can impact the acceptance of technology by end-consumers, especially when the media reports high profile stories of security services collecting personal information without people's knowledge.

Sharing data, be it through one's personal or home devices or in a public space, is becoming ubiquitous. Technology enables us as individuals and companies through data harvesting to increasingly collect a wide range of personal data. For example, detailed information might be gathered on when residents are at home, whether they are watching TV, or cooking dinner. It is however unclear how privacy is or should be protected when data is transferred across multiple systems and technology owners. This includes the protection of personal information (e.g., social identity, health information, etc.), personal communication (e.g., emails, text messages), and personal behavior (e.g., information on daily routines) [2]. These issues become even more pronounced when information is exchanged not just between an end-user and the government or a private company, but between multiple users in real time and for commercial gain.

Another area of regulation that is becoming more complex is the issue of security and liability. Questions around who is responsible for security breaches or other accidents when machine to machine communication fails need to be resolved. An example of where this will become increasingly crucial is the deployment of auto-mated vehicles on city streets. If an automated vehicle drives into a parking lot based on geospatial data, but crashes into another car, responsibility could be with the manufacturer of the automated vehicle, the software producer, or the data provider.

3.3   Economic Challenges

A city-wide ICT infrastructure and the overarching software to achieve it, whilst requiring significant investment, are nonetheless often an order of magnitude smaller in cost terms than the main city infrastructures such as highways, transport systems and fixed utility systems. However, these systems have struggled to attract funding. At its core, the value proposition for smart and cognitive cities has still not been made effectively and investments have not been packaged into manageable chunks with clear outcomes related to key business cases. In part, this stems from the fact that with few examples of city-wide and sector-integrated roll-out of technology infrastructure, it is still difficult to justify investment following a traditional methodology.

Technology companies looking for new market opportunities have contributed to this challenge without offering any suitable alternatives [5, 18]. IBM, Siemens, HP, Cisco, Toshiba, Hitachi, and others have smart city research and product development divisions that target cities with the opportunity for direct investment in their product lines. This model has not demonstrated clear social or financial value to the cities and has not worked as city administrators do not have the capital required or the desire to lock themselves into costly service contracts over a long timeframe.

It is noted that there are a few initiatives that are invested in by the cities' own operational budgets [generally via a Public Private Partnership model (PPP)] if a sufficient return on investment is demonstrated. An example is PARK Smart in New York City,[2] which combines real-time parking information via web and smartphone with progressive meter rates. Occasionally, such smart systems can be incorporated as part of major upgrades of utilities. In the wake of Hurricane Sandy, local utility companies received permission to undertake large-scale strengthening of their networks and were allowed to set aside 10 % of the funding to create much smarter networks [22].

Most smart city initiatives in Europe and North America are however financed by small scale research funding and pro bono investment on small pilots by technology companies hoping to demonstrate proof of concept and thus to progress to larger and better paid contracts. Research bodies such as the EU Horizon 2020 program or the UK's Technology Strategy Board will continue to invest in the next round of smart (or eventually cognitive) city ideas as it supports the broader macro-economic opportunity by fostering the next generation of creative and digital companies.

However, for larger-scale implementation of smart and cognitive systems, the cities still face major challenges: City administrations do not always understand the potential value of the digital opportunity and face therefore difficulties in assessing or justifying the up-front and on-going operational costs needed for the investment. At the same time, utility companies do not always want to invest in smart infrastructure because, while it could save consumers money, they perceive a reduced level of return on the investment in assets already installed and operational, and returning a good level of profitability.

3.4 Social Challenges

There are several social issues that cities face when dealing with new technologies. For instance, shared economy applications such as the car service Uber or the

---

[2]PARK Smart NYC: http://www.nyc.gov/html/dot/html/motorist/parksmart.shtml [accessed 02.10. 15].

short-term rental application Airbnb are beginning to restructure and challenge the way we traditional move and inhabit our cities.

Citizens are the ultimate stakeholders in the city and questions such as who owns personal information and what can be done with it become crucial to address. If, in a cognitive city, citizens are persuaded (either consciously or unconsciously) to provide information about the way they behave in the urban environment, there will be a need for clear and transparent regulations around (data) privacy and security [13]. This is especially important as citizens often do not have the knowledge and understanding of how they are interacting with technology and what the benefits and drawbacks might be. The success and optimization of any technology systems will rely on the ultimate user actually using it in the way it was intended. Initiatives that are solely technology-oriented and do not consider the needs of end-users will more than likely fail.

In addition, questions of inequality and segregation are rarely addressed in technology debates. When technology is becoming ubiquitous, there is a danger that those without the educational, financial, or technological capital to participate in the transformation are left behind. Often these are the most vulnerable segments of the population such as the elderly or low-income communities. This is illustrated by two examples. Crowdsourcing applications such as SeeClickFix[3] let citizens report problems through their smartphones or online; governments can then use this information to address the problems. If city governments start only to address (or prioritize) those problems reported through the app, those neighborhoods that do not have an active 'online' population will be left out Townsend [20]. Similarly, those businesses that do not have an online presence are in danger of becoming obsolete in a world where decisions on choice of business are based on search ratings. There is thus a real danger that technology reinforces existing social inequalities and spatial segregation—unless these issues are specifically addressed.

3.5 Technological Challenges

Traditional infrastructure such as electricity grids or transport systems tends to have a long lifecycle in comparison to the short lifecycles of ICT systems and even shorter lifecycles of software applications. Infrastructure networks need to be adaptable to support changes both in peak demand and in the implementation of new systems over time that might be faster and more efficient than legacy systems. Ideally, infrastructure networks should have the ability to 'plug and play' new technologies as they become available. It is however not just the engineering robustness and flexibility that has to be considered. Cities are often locked into these legacy systems with well-established contractual, commercial, and legal structures that deter innovation and adoption of new technologies.

---

[3]See ClickFix: http://en.seeclickfix.com/ [accessed 02.10.15].

At the same time, technology systems have traditionally been developed to be deployed in functional silos, with proprietary hardware and protocols that do not allow ease of integration with other systems. In a similar vein to the issues with legacy infrastructure, the well-established commercial, contractual, and legal structures also hinder the adoption of new technology systems. At the rate that technology is currently changing, this has (or will have) a major implication in relation to the commissioning and adoption of new systems.

There are thus a number of challenges to deliver a cognitive city in a physical context, in addition to the lack of common interfaces and operating systems, there is the problem of the capacity and capability to collect, store, and analyze the vast amount of data that is transferred every second. Currently, standards for interoperability for smart or cognitive city systems are still in the early phases and even the way data is collected, stored and presented may vary between different departments within a city. If a simple step such as a comprehensive data management framework can be implemented, monitored, and governed by an appropriate body within the city's municipal organization, cost savings in processing, storage and database management can be achieved. Additionally, improved access to data collated from multiple sources can provide a far richer picture of what is happening at a system and citizen level, enabling more informed decision making through real-time learning and adaption.

Finally, cities not only need to adopt new technologies that help them collect data, they also need to be able to easily access and analyze the data in real time to enhance the learning and memory creation through which improved systems responses emerge. This requires a certain degree of standardization of data collection, storage, display, and reporting across city systems—and preferably within a common and aligned time-frame. Operating systems that formalize standard schemas and adhere to an open and transparent data management policy can help the city with such standardization. There are a small number of operating systems that have been successfully developed as pilot schemes; Living PlanIT's Urban Operating System (UOS)[4] and IBM's Intelligent Operations Center[5] are two examples. Neither, however, has been deployed at city scale covering the full range of city departments and city systems.

Having explored the political, regulatory, economic, social, and technological challenges, perhaps the most important challenge of all is the need to understand the city as a complex, adaptive system. Too often technologies are deployed in an un-coordinated way, trying to solve one problem but ignoring many other interrelated issues. Or they tend to solely focus on the advancement of technology and the system efficiency that they bring. Or they focus on the possibility of collecting data without being able to effectively process and use it. One way or another there is the danger of ambivalence and 'death by data'.

---

[4]Planit UOS: http://www.living-planit.com/ [accessed 02.10.15].

[5]IBM Intelligent Operations Center: http://www-03.ibm.com/software/products/en/intelligent-operations-center [accessed 02.10.15].

4 Successful Cities—Technology Is Only One Component

The success of cities is context- and path-dependent. While there have been attempts to build 'smart' cities from scratch, like Masdar City in the United Arab Emirates or PlanIT Valley in Portugal, most people will not live in these newly built cities, but continue to live in already established cities. Successful smart or cognitive cities are thus cities that can leverage technology to enhance the living standards of their incumbent and future populations and the performance of their existing economies. Cities are foremost places for human interaction: to do business, to live, and to enjoy the economic, social, and cultural benefits that dense human settlements provide. Technology is thus not the end goal, but a mean to address the various challenges outlined previously.

4.1 Livable Cities

Successful cities enhance the quality of urban living by providing education, job, and business opportunities, support social well-being through its built environment and services, and minimize the environmental impact. They also provide open and transparent government and public service and infrastructure systems that are affordable and reliable. In short, they are livable cities (see Fig. 1).

Technology can enhance these aspects of the city, but in order to harness the full benefits of technology, cities and their governments need to understand the opportunities and risks. And they need to evaluate their use of technology on the basis of how it creates added value through memory creation and learning to improve the environmental, social, and economic aspects of a city. This requires integrated thinking that connects the various physical and social infrastructure systems of the city to deliver better services with more value for citizens.

Cities such as Amsterdam and Copenhagen have developed strategies that broadly demonstrate how smart cities *could* evolve. The same cannot be said for cognitive cities. Copenhagen however, aims to be the first carbon neutral capital by 2025 (stateofgreen.com) and has recently announced its partnership with Hitachi Consulting to develop a citywide urban operating system. This will allow the city to improve its systems through feedback loops that measure changes in the system in real-time, therefore enhancing the machine-to-human and machine-to-machine learning. The city will use technology and innovation not just to advance its technology sector, but to apply a system-wide thinking that benefits its citizens and the city's livability. The city has made great advancement in waste reduction through educational outreach, new composting and recycling schemes, and an upgrading of its waste-to-energy plants. With landfill from waste reduced to 1.8 and 98 % coverage by district heating, its results are impressive compared to most other cities.

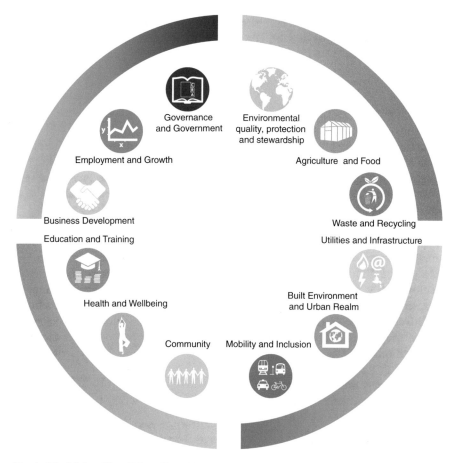

Fig. 1 The Living City. (©BuroHappold)

Similarly, Amsterdam Smart City (amsterdamsmartcity.com) has an ambitious agenda of thinking across sectors. A pilot project called Vehicle2Grid,[6] currently underway, allows residents to control their own energy supply. An open data platform connects households with car batteries or electric charging stations and a smart grid. A smart phone app then allows residents to decide how they want to use the locally produced energy from solar panels or wind farms. The energy can be fed into the smart grid, used immediately to drive a car or run household appliances, or stored in the battery of an electric car. This strategy demonstrates the benefits in terms of greater control and choice for consumers over their energy consumption and provides resilience in energy supply.

---

[6]Vehicle2Grid: http://www.amsterdamvehicle2grid.nl/ [accessed 02.10.15].

Both examples mentioned here demonstrate how technology is just one component of many (e.g., regulatory, educational, social etc.) that create a successful urban environment.

4.2 Engaged Communities

A successful city is a city that encourages its residents to engage. This does not mean that everyone needs to be involved at every level of decision making or activity. Instead, it means that city governments work to get the buy-in for new initiatives from stakeholders and allows them to contribute to the development of technology strategies.

Engaging with residents and communities through communicating the benefits to them is crucial for the deployment of technology strategies; even more so if individuals can be encouraged to provide data for the improvement of urban systems. In the US, for example, the introduction of smart meters has been highly controversial with users raising concerns over cost, privacy, and health. An early pilot scheme in Boulder, Colorado failed to inform its customers of the benefits to be derived from the newly installed smart meters and the opportunity to check one's energy use online. Public engagement and education campaigns are therefore crucial in order to successfully leverage the benefits of technology improvements.

Allowing individuals and communities to contribute to the implementation of technology may mean encouraging citizens to use the opportunity by creating open data policies and offering educational programs. This enables individuals and communities to create new ways to experience and improve cities, services, and information. It also creates a vibrant community of creative thinkers and entrepreneurs. An interesting example of achieving this can be found in Chicago. The City of Chicago, together with its partners MacArthur Foundation and the Chicago Community Trust, has funded Smart Chicago, a civic organization that is devoted to increasing access to the Internet amongst Chicagoans, improving skills required to use the Internet, and developing products from data to enhance the quality of life in the city (smartchicagocollaborative.org). It is through these early adoptions of technology that cognitive cities can emerge where technology is used to learn, create, and adapt.

4.3 High Capacity, Integrated, and Resilient ICT Infrastructure

Without a high capacity and resilient ICT infrastructure, cities will not be able to fully harness the benefits of technology. It also needs to be coordinated with and integrated into other infrastructure systems and networks across the city both physically and virtually. As Moss-Kanter and Litow [14: 2] put it: "Infusing intelligence into each sub system of a city, one by one—transport, energy, education, health care, building, physical infrastructure, food, water, public safety, et cetera, is not enough to become a smarter city. A smarter city should be viewed as an organic whole—as a network, a linked system." With regards to ICT infrastructure, this means a network of sub systems, interconnected and integrated through open standards, shared infrastructure and common protocols.

South Korea is probably most advanced in this regards. It has invested enormous public funds into ICT infrastructure and relaxed regulations to provide a competitive platform for telecommunication service providers. Today, Seoul has a free public Wi-Fi that offers the world's fastest Internet speeds. South Korea however has not stopped there and has just announced the upgrade of its mobile infrastructure with the aim to increase its Internet speed by 1000 times. As a consequence, South Korea increasingly competes with Silicon Valley for investment into the information technology industry [25].

Cities are complex systems of systems that are the outcomes of various social, economic, and institutional relationships. Technology can play a significant role in improving the efficiency of these relationships, therefore contributing to resource maximization and waste minimization. Successful cities require more than simply the adoption of smart and cognitive devices and systems. This also needs to be considered when implementing a roadmap for adopting technology.

5 Roadmap to Adopting Technology

While there are no 'one-size-fits-all' solutions to adopting technology, there are a number of key ingredients which will facilitate the creation of smart and cognitive cities. These can be broadly divided into five categories: vision, leadership and organization, citizen engagement, financing, and technology (see Fig. 2).

5.1 Create a Shared Vision

The first step in adopting technology and building cognitive cities is the creation of a shared vision. Questions such as what, why, and how need to be answered. The process of developing a vision includes the evaluation of existing conditions, the assessment of future opportunities, and finally the development of a framework or plan that establishes the vision, goals, strategies, targets, and key performance indicators. In developing this plan, those responsible for governing cities need to understand what they want to achieve and what the required return on their investment is. Plans should not be considered as set in stone and need to be flexible and adaptable to new circumstances and innovations. It is thus essential that these plans are reviewed and, where necessary, updated on a regular basis. In summary, plans should be ambitious, achievable, accessible, accountable, and adaptable.

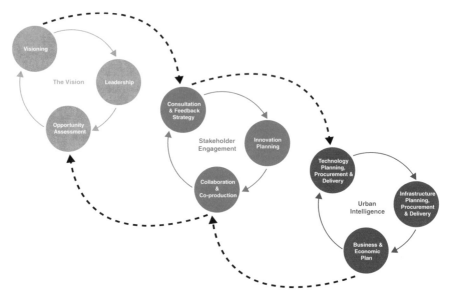

Fig. 2 Roadmap—vision, engagement, intelligence. (©BuroHappold)

Cities with strong smart city aspirations (no city has officially declared any cognitive city aspirations yet) have developed policy frameworks setting out their vision, key performance indicators, and strategies for delivery. New York was one of the earliest proponents of promoting technology to improve the city. In its 2011 "Roadmap for the Digital City", it outlined four goals, namely: to ensure that all New Yorkers can access the Internet, to support open government policies to unlock public information and increase transparency, to enable citizen-centric and collaborative governance through digital tools including social media engagement and, finally, to support a digital media sector [3].

More recently developed, the Smart London Plan considers the role of digital technology in planning for London's future population growth challenges. It has a clear vision statement aiming to "use the creative power of new technologies to serve London and improve Londoners' lives" and is set within an overarching framework of the Mayor's 2020 vision. It outlines a set of high-level goals from placing "Londoners at the core" to "enable London to adapt and grow". Each of the goals has a number of strategies and measures of success. Taking the goal of enabling London to adapt and grow, the plan outlines six strategies, including the "promotion of smart grid technologies" to help meet the increased demand for electricity and "experiment with new ways of reducing light freight" to tackle the rise in congestion and pollution caused by increased e-commerce. Indicators to measure progress include "work towards a reduction of greenhouse gas emissions to reach 40 % below 1990s levels by 2020" [4]. Whilst both examples demonstrate forward thinking, they could go still further by placing greater emphasis on the benefit of integrating data across systems and creating targets and metrics that support this.

Fundamentally, any vision should include input from stakeholders, and ensure it is transparent, accountable, and accessible. As highlighted earlier, citizens are key to the success of any cognitive city initiative and, if they are not engaged, the benefits of introducing technology will not be realized in full.

5.2 Establish Strong Leadership, Organizational Structure, and Regulatory Framework

The adoption of technology will need more than a vision and a plan. Implementation requires a strong leadership that is committed to the long term development of the city and the wellbeing of its citizens. Structures of city administrations vary; in some cities strong mayors will lead progress towards technology solutions, in others a chief technology officer will be responsible. In addition, the role of city departments and their capacity and capability to deliver, operate, and maintain technology solutions will also be key. This is often a challenge for local authorities as they have to compete with the private sector for the best employees. It should be recognized that leaders and staff often change and there is always the risk that last year's priorities and efforts become this year's abandoned follies. This needs to be considered and measures put in place to ensure that long term strategies and targets have broad support and are deliverable. Creating a culture to think long term will also set the attitude regarding appropriate timetables for delivery, return on investment, and the sustainability agenda. Finally, the city needs to develop a regulatory and legal framework around data privacy, data security, and liability.

New York demonstrated strong leadership in developing the digital roadmap, with Mayor Bloomberg spearheading the early adoption of open data and online governance services. Mayor Bloomberg also managed to establish an organization that carries forward the legacy. New York City's Department of Information Technology and Telecommunications is charged with the modernization of IT infrastructure, the transparency of city government through its open data portal, and the development of innovative solution to improve the delivery of city services. It is one of the largest cross-sector technology city departments with approximately 1200 employees, an operating budget of $375 million, and a capital budget of $1 billion over four years [17].

Smaller cities will not have the resources to establish such departments. However, they may still be able to develop a policy framework and hire a chief digital officer to operate across departments to coordinate and integrate otherwise channeled efforts. Townsend [20] goes further in recommending that cities without the capacity to build their own software applications should use the many bottom-up initiatives developed by start-ups across the world. They can work with

third-party partners such as academic institutions and private companies to integrate
some of the innovative solutions.

5.3 Engage Stakeholders Through the Process

The key aim of the cognitive city has to be to create benefits for citizens and to
improve their quality of life through use of technology. For any city administration
this takes on a number of aspects: engaging residents in the process of helping to
develop strategies, creating physical and virtual space to provide input and feedback
to planning proposals, supporting open data, encouraging entrepreneurship via
incubators and shared working spaces, and ensuring buy-in through public
outreach.

Singapore has developed one of the most advanced water treatment systems to
recycle sewage water and purify it into ultra clean drinking water. Without gaining
the buy-in of Singaporeans to drink this water, the initiative could have been a
complete failure. Singapore's national water agency staged a large-scale public
education and awareness campaign to create acceptance and demand for the treated
water and has thereby established a successful outcome.

Gamification is one area that cities are now exploring to enhance citizen
engagement. Governments apply the principles of computer gaming to encourage
people to learn online, encourage participation in planning discussions (for example
about planned bus routes), and volunteer their time. In Stockholm, for example, the
city administration used gamification to encourage safer driving. A speed camera
entered the details of those who obey to speed limit into a lottery funded by those
who are fined for traffic violations. The impact was that traffic speed decreased by
22 % [23]. Software companies like MetroQuest are increasingly being employed
by cities to help create portals that allow the public to interact and take part in
making decisions on projects that have a citizen focus.

5.4 Develop Business Case, Financing Mechanisms, and Procurement Models

Smart and cognitive systems require a large number of technology-based infrastructure interventions to be deployed. The integration of these has been difficult to
realize at a large scale, primarily due to poor business case planning and market
engagement but also due to the immaturity of integrated solutions. Too often, cities
have passively waited for other actors to formulate propositions, most of which
have been rejected as inappropriate to city needs. In the future, the city will need to
play a more active role in developing propositions—even where the investment
participation of other stakeholders will be required.

Investment in upgrading infrastructure to become smart or cognitive usually requires high upfront costs with long-term and potentially volatile returns that may or may not appeal to private investors and some of which may not be of a financial nature. To resolve this, it will most likely require a combination of public and private sector funding. It also suggests that a realistic articulation of private sector benefits (internal rate of return) and public sector benefits (socio-economic return) will need to be at the center of decision-making.

The need for a clearer understanding of long-term benefits will require cities and investors in cities to model scenarios before making any decision on large-scale investments. This will help with the engagement of the private sector by demonstrating the commercial and financial viability (or not, as the case may be). Any financial model should identify market potential. Typical components of one will include:

- *Required rate of return*: Return that is at least as good (and ideally higher) as an investment in the financial market with similar risks.
- *Debt and equity ratio*: Lower debt ratio normally indicates a less risky investment due to the volatility of interest rates; it will however depend on the type of solution proposed and its market drivers.
- *Costs*: Operational (OPEX) (the ongoing cost for running a system such as an electric grid with sensors and meters) and capital (CAPEX) (the capital upfront investment for the installation and adoption of a system) expenditures as well as additional charges (e.g., licenses, management charges, operating costs).
- *Revenues*: Income from user fees and sales or rental of infrastructure and equipment. This is typically linked to an appropriate length of contract.
- *Discount rate*: Rate required to calculate the present value of future cash flow. Higher discount rates should be used to adjust for risk and opportunity cost associated with specific forms of digital solution, particularly those that are relatively untested in the market.

It is important to note that technologies can have a positive (or negative) impact beyond the narrowly defined return on investment. In some cases, the return on investment might be marginal or neutral at best, but its wider economic impact (e.g., employment creation, health benefit etc.) may be substantial. Parties to such a potential investment therefore need to undertake robust socio-economic impact assessments in order to understand the wider benefits that may accrue from technology solutions. Typical components of a socio-economic impact assessment include:

- *Direct impacts*: Growth/decline related to project activity and/or construction phase.
- *Indirect impacts*: Growth/decline related to supply and wider value chains.
- *Induced impacts*: Shifts in income emanating from direct and indirect impacts.

- *Multipliers*: Overall growth based on indirect and induced impacts.
- *Socio-economic impacts*: Economic competitiveness, health and well-being, marketability, environmental impact, etc.

Financial models and socio-economic impact assessments will help identify appropriate procurement models to engage the market. Different types of investors will have different risk and return strategies and will want to buy in (or sell out) at different points in the development process. City administrators therefore need to create and manage procurement models that provide a 'win-win' for the city, consumer, and provider over appropriate time periods.

5.5   Technology Identification and Selection

In a fast changing sector such as technology, it is key to identify and select the appropriate technology systems to ensure that the vision and goals are met within an acceptable budget. Following on from the procurement strategy, it is important that early engagement of the broader technology market takes place in order to identify and assess the alternative solutions that can be delivered to meet outcomes specified by those promoting the infrastructure and support structure. This is summarized through the following broad steps.

There needs to be a consistent market scanning and knowledge sharing across cities. The current technology market is vast and trying to undertake a comprehensive procurement exercise without the benefit of a thorough understanding of the market offer and its performance increases the risk of selecting an inappropriate technology solution or missing a cost-effective response. Cities should consider sharing their experience and knowledge on a continuing basis to ensure that best practice and knowledge is propagated.

Evaluation criteria need to be developed to support the procurement process for technology selection. These could include technological feasibility, regulatory feasibility, fit with context, interoperability, business case, and support of quality of life conditions.

Once the broad business and system requirements of the technology solution are understood, an expression of interest (EOI) to vet the market's appetite should be issued to the market. This allows those responsible for procurement to not only broadly understand what the market can offer, but it also provides information on the maturity of the solutions, the economic and financial standing of the market to deliver the solutions, and the appetite of the market to engage with different procurement models.

In parallel to gathering early expression of interest proposals from potential developers, there should be consideration of the impact of any proposed technology on the infrastructure networks (existing and proposed) required to support it. This review should cover not only the spatial impact but also the impact that shared data may have on organizational and commercial structures.

Proposals should be shortlisted based upon the evaluation criteria before advancing to the next stage. The next stage should request more detailed proposals (technology and commercial) of the solution and have greater definition of the required specification and the business, system and data requirements so as to help ensure that the technology fulfils expectations. This stage is commonly referred to as the RFP (Request for Proposals). It will lead to a final recommendation, contractual and commercial negotiations, and contract signing (with an optional back-up proposal in case the preferred commercial deal falls through).

6 Conclusion

In the twentieth century, many cities were re-designed to accommodate car travel, resulting in a long-term legacy that has negatively impacted the environment and health and well-being of citizens. Whilst smart and cognitive city technologies may seem less obtrusive than highways, their legacy will similarly impact our cities and citizens for decades to come. Having outlined the key steps to overcome the current challenges to the adoption of technology, we ultimately argue for careful consideration of how we harness technology to fully realize the opportunities of it.

Technology has enormous potential to help tackle the challenges of our rapidly growing cities. It can improve decision making based on real-time data, memory creation, and learning, enabling cities to provide better services to increase quality of life, and to do so in a more financial and natural resource efficient way. There is however a real risk that technology could do more harm than good. Technology can easily become an excuse to continue our resource-intensive consumer behavior with the argument that technology will, eventually, solve all problems. The enthusiasm around self-driving cars is a case in point. It is argued that self-driving cars will eliminate any traffic accidents and improve efficiency. However it will not necessarily solve the problem of congestion of our car-oriented, unhealthy lifestyles.

To make the most of technological advancement, cities and their administrators need to focus on the end user needs and on long-term outcomes. A successful city actively enhances the quality of life for its citizens using technology as an enabler. There is thus an urgent need to understand the impact of technology. Whilst there is an abundance of literature on smart city initiatives, there is little research being undertaken on the impacts of these initiatives on individuals, neighborhoods, and broader city communities. Academia could play a critical role in bringing insight into the consequences of smart and cognitive cities. Ultimately, better analysis and better delivery models are required to be developed for technology solutions.

Successful cities are places where people want to live, work, and play. Being conscious of this, city administrators should develop a roadmap that makes their priorities clear. And whilst technology should be afforded an important place in their considerations, it is just one infrastructure layer among many other interrelated components, both physical and social that makes our cities successful.

Acknowledgments The authors would like to thank their colleagues at BuroHappold Andrew Comer, Paul Goff, and Lawrie Robertson for their contributions and comments.

References

1. Avent, R.: The third great wave. The Economist, Oct. 4th (2014)
2. Bartoli, A., et al.: On the ineffectiveness of today's privacy regulations for secure smart city networks. In: Proceedings of third IEEE International Conference on Smart Grid Commuications, Tainan City Taiwan, 5–8 Nov 2012
3. City of New York: Road Map for the Digital City. Achieving New York City's Digital Future (2011)
4. Greater London Authority: Smart London Plan (n.d.)
5. Greenfield, A.: Against the smart city. Do projects, New York City (2013)
6. IBM: IBM and City of Portland Collaborate to Build a Smarter City. New release, 09 Aug 2011, Portland, Oregon. http://www-03.ibm.com/press/us/en/pressrelease/35206.wss (2011). Accessed 24 Jun 2015
7. IPCC: Climate Change 2014: Mitigation of Climate Change—Chapter 12 Human Settlements, Infrastructure and Spatial Planning. Potsdam, IPCC—Working Group III (2014)
8. Kaltenrieder, P., et al.: Enhancing multidirectional communication for cognitive cities. In: eDemocracy & eGovernment (ICEDEG), 2015 Second International Conference, Quito, 8–10 Apr 2015 (2015)
9. Kaltenrieder, P., et al.: Applying the fuzzy analytical hierarchy process in cognitive cities. In: Conference Proceedings of the International Conference on Theory and Practice of Electronic Governance, pp. 259–262 (2014)
10. Khansari, N., Mostasharti, A., Mansouri, M.: Impacting sustainable behaviour and planning in smart city. Int. J. Sustain. Land Use Urban Planning **1**(2), 46–61 (2013)
11. LSE Cities: Cities and the New Climate Economy: the transformative role of global urban growth. In: NCE Cities—Paper 01. The Global Commission on the Economy and Climate, Nov 2014
12. Mayer-Schöneberger, V., Cukier, K.: Big data: a revolution that will transform how we live, work, and think. Hachette, London (2013)
13. Mostashari, A., Arnold, F., Mansouri, M., Finger, M.: Cognitive cities and intelligent urban governernance. Network Ind. Q. **13**(3), 4–7 (2011)
14. Moss-Kanter, R., Litow, S.: Informed and interconnected: a manifesto for smarter cities. Harward Business School Working Paper 09–141 (2009)
15. Moyser, R.: Planning for smart cities in the UK. http://www.burohappold.com/blog/post/planning-for-smart-cities-in-the-uk-2179 (2013). Accessed 2 Jun 2015
16. Neirotti, P., De Marco, A., Cagliano, A.C., Mangano, G., Scorrano, F.: Current trends in smart city initiatives: some stylised facts. Cities **38**, 25–36 (2014)
17. NYC Department of Information Technology & Telecommunications (NYC DoITT): Strategic Plan 2015–2017 (2015)
18. Robinson, R.: Creating successful Smart Cities in 2014 will be an economic, financial and political challenge, not an engineering accomplishment. The Urban Technologist. People. Place. Technology March 2, 2014. http://theurbantechnologist.com/2014/03/02/creating-successful-smart-cities-in-2014-will-be-an-economic-financial-and-political-challenge-not-an-engineering-accomplishment/ (2014). Accessed 2 Jun 2015
19. The Climate Group, Arup, Accenture, Horizon: Information Marketplaces. The New Economics of Cities (2011)
20. Townsend, A.: Smart Cities: Big Data, Civic Hackers, and the Quest for a New Utopia. W. W. Norton & Company (2013)
21. UN (United Nations): World Urbanization Prospects. 2014 Revision. United Nations (2014)

22. Warshay: Upgrading the Grid—How to modernize America's electrical infrastructure. Foreign Affairs March/April 2015 (2015)
23. Wood, C.: Gameification: governments use gaming principles to get citizens involved. Government Technology, 28 Aug 2013. http://www.govtech.com/local/Gamification-Governments-Use-Gaming-Principles-to-Get-Citizens-Involved.html (2013). Accessed 08 Jul 2015
24. World Bank: What a Waste: A Global Review of Solid Waste Management. World Bank (2012)
25. Wortham, J.: What Silicon Valley can learn from Seoul. New York Times June 2, 2015. http://www.nytimes.com/2015/06/07/magazine/what-silicon-valley-can-learn-from-seoul.html?smid=nytcore-iphone-share&smprod=nytcore-iphone&_r=0 (2015). Accessed 4 Jun 2015

Maturity Model for Cognitive Cities

Three Case Studies

Luis Terán, Aigul Kaskina and Andreas Meier

Abstract The key mission of cognitive cities is to bring into symbiotic cooperation the government, citizens, and the business. To achieve this, cutting-edge information technologies are used to build smarter cities (cognitive cities) in which administrations stand as principal providers for smart services to citizens, constructed communication between business/industry, and more efficient city governance. As a result, citizens, businesses and government are empowered with information that eases the relations between these three parties and the inclusion of cognitive processes for decision-making. This chapter focuses on the interaction between administrations and citizens. It highlights an eGovernment framework through which eEmpowerment of citizens can be achieved via the promotion of citizens' participation. With the help of information technologies, such as geolocation, the Internet of things, and open source software and hardware, among others, smooth communication between government and citizens can be established, thus facilitating the decision-making processes in cognitive cities. The illustration of three different applications of intelligent agents shows how the eParticipation of citizens can be increased. The first case explains how collaborative working environments are used in the public sector for a collaborative legislation process, to support people in a virtual space in a time- and place-independent manner in the creation of an organic law. The second application illustrates how

L. Terán (✉) · A. Kaskina · A. Meier
Department of Informatics, University of Fribourg, CH-1700 Boulevard de Pérolles 90,
Fribourg, Switzerland
e-mail: luis.teran@unifr.ch; lfteran1@espe.edu.ec

A. Kaskina
e-mail: aigul.kaskina@unifr.ch

A. Meier
e-mail: andreas.meier@unifr.ch

L. Terán
Universidad de Las Fuerzas Armadas (ESPE), General Rumi–ahui
S/N, Sector Santa Clara - Valle de los Chillos, Sangolquí, Ecuador

© Springer International Publishing Switzerland 2016   37
E. Portmann and M. Finger (eds.), *Towards Cognitive Cities*,
Studies in Systems, Decision and Control 63, DOI 10.1007/978-3-319-33798-2_3

eDemocracy processes can be facilitated with the help of a so-called social voting advice application, which takes into consideration trust values among citizens and governmental representatives. Finally, the third application highlights how civic participation can be enhanced through building communities of interest and communities of practice within the *SmartParticipation* platform.

1 eGovernment Framework Applied on Cognitive Cities

The concept of the "cognitive city" represents an information-centric community, and entails the existence of learning, memory creating, and experience retrieval that continuously improve urban governance [1]. The cognitive city is a paradigm of a smart community that takes advantage of information technology trends, such as artificial intelligence, natural language processing, and data mining including a human cognition analysis. It is build upon the cognitive system that learns and adapts its behavior based on past experiences and is able to sense, understand and respond to changes in its environment [1]. In the cognitive city, the citizen is a vital element for efficient and effective functioning since s/he is the principal source of data generation, as well as the active consumer of the urban information. The eGovernment of the cognitive city is tightly connected to citizens' civic participation through means of smart applications, social networks, web-applications and other smart devices. Thus, cognitive cities lead to eEmpowerment, where citizens are facilitated with smart and intelligent tools that accelerate decision-making processes between citizens and governmental institutions.

According to Meier [2], the term "electronic government", or eGovernment, implies the simplification and execution of information, communication, and interchange processes within and between governmental institutions, and also between the governmental institutions and citizens or organizations. It places electronic governmental services and public transactions available in a time- and place- independent way to citizens (e.g., taxation, social facilities, employment service, social security, official ID cards, health services, etc.) and companies (e.g., taxes, company start-ups, statistical offices, customs declaration, environmental performance, public procurement, etc.). In this chapter, the Electronic Government Framework proposed by Meier [2] is used to define how eEmpowerment can be reached. Figure 1 shows the eGovernment Framework that includes 3 levels: Level I—Information and Communication, Level II—Production, and Level III—Participation.

This chapter focuses in the third level, eParticipation, of the eGovernment Framework, taking into account the 3 elements presented: eCollaboration, eDemocracy, and eCommunities. The eGovernment Framework should be applied for the development of cognitive cities that use information and communication technologies (ICT) to enhance quality, performance, and availability of services. They aim to reduce costs and resource usage by engaging citizens to profit from the

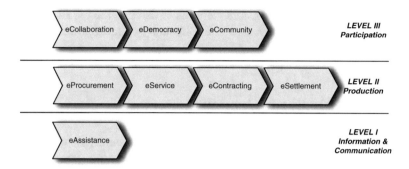

Fig. 1 eGovernment framework. Adapted from Meier [2]

development of digital solutions, including government services such as transport and traffic management, energy, health care, and water and waste, among others. In the work of Schaffers et al. [3], different Internet technologies and components for Smart Cities are summarized in Table 1.

Table 1 shows the importance of the use of scalable systems to provide new eServices in an innovation ecosystem. Additionally, collaboration and participation from citizens is a key factor for the success in the implementation and execution of Smart City projects.

This chapter presents practical applications of eCollaboration, eDemocracy, and eCommunities processes based on eGovernment framework [2] toward the eEmpowerment of citizens in the cognitive cities. In the next section, a five-level maturity model toward eEmpowerment is presented and used for the case of smarter cities (cognitive cities) with the use of intelligent agents. Additionally, a study case

Table 1 Media internet technologies and components for smart cities. Adapted from Schaffers et al. [3]

Solutions and challenges	Short term	Mid term	Long term
Content management tools	Media internet technologies	Scalable multimedia compression and transmission	Immersive multimedia
Collaboration tools	Crowd-based location content; augmented reality tools	Content and context fusion technologies	Intelligent content objects; large scale ontologies and semantic content
Cloud services and software components	City-based clouds	Open and federated content platforms	Cloud-based fully connected city
Smart systems based on internet of things	Smart power management portable systems	Smart systems enabling integrated solutions (e.g., health and care)	Software agents and advanced sensor fusion; telepresence

is presented for each of the proposed elements to describe the use of collaborative environments, recommender systems, and community-building processes within the scope of the maturity model.

2 Maturity Model for Cognitive Cities

The need of citizens and other stakeholders for a free democratic debate, participation, and the right to be involved in the decision-making process, has been highlighted by different theorists. The use of ICT has opened new channels for free discussion of political issues, and day by day it is moving away from traditional media such as TV, radio, mail, and newspapers. *eParticipation* has been addressed more often in the academia, and is an emerging and growing research area aimed at enhancing citizens' participation to providing a fair and efficient society with the support of administrations by the use of cutting-edge technology.

In this chapter the model proposed in the work of Terán [4], inspired by Tambouris et al. [5] is presented. It includes the concepts of Web 2.0 in order to provide community-building processes and discussion between citizens and the administration. It consists of five levels: *eInforming*, *eConsulting*, *eDiscussion*, *eParticipation*, and *eEmpowerment*. Each of these levels is described in more detail as follows:

2.1   Level I—eInforming

This first level uses a unidirectional (top-down) channel to provide citizens with relevant information about different policies, projects, and news, among others. At this stage, citizens are informed by the administration; no interaction, participation, or decision is present. An example of this level is the Web portals of city administrations.

2.2   Level II—eConsulting

This corresponds to the second level and uses a bi-directional channel giving the authorities the ability to collect feedback from citizens. At this stage, citizens are consulted by the government and minimal interaction is present. Nevertheless, neither participation nor decision is present. An example of this level corresponds to public opinion polls, where the administrations aims to represent the opinions of citizens by conducting a series of questions.

2.3 Level III—eDiscussion

This corresponds to the third level and uses a bi-directional channel. At this level, administrations provide citizens with the ability to establish discussion channels and create virtual communities by building citizen communication centers is defined. Public project ideas and plans can be discussed and commented on, taking advantage of specialized groups (communities) in order to promote the opinion-forming process. At this stage, citizens are able to establish communication channels. Nevertheless, neither participation nor decision is yet present. An example of this level is presented in the work of Palen and Liu [6].

2.4 Level IV—eParticipation

This corresponds to the fourth level and uses a bi-directional channel. It provides citizens with the ability to collaborate on public projects and developing bases for decision-making. At this stage, citizens are able to establish much bigger communication channels, which include more capabilities such as working collaboratively to enhance participation. The first steps toward empowerment are taken here. The following applications are considered elements on a collaborative working environment (CWE): e-mail, instant messaging, application sharing, video conferencing, collaborative workspace, document management and version control system, task and workflow management, Wiki group or community effort to edit wiki pages, and blogging systems, among others. An example of collaborative e-government is presented in the work of Ae Chun et al. [7].

2.5 Level V—eEmpowerment

This corresponds to the fifth level and uses a bi-directional channel. It places the final decision in the hands of the citizens, thus implementing what they have decided. At this stage, citizens are empowered, as the communication channels are much bigger and include new and better capabilities towards empowerment. In their work, Wang and Ruhe [8] mention the importance of cognitive processes in the decision-making. They propose a set of decision strategies have been proposed from different angles and application domains. To visualize an application of eEmpowerment, refer to the example of voting processes in a digital participative budget is presented in the work of de Souza and Maciel [9].

It is important to point out that, from the processes in Figs. 2, 3, 4, 5 and 6, the comparative size of the administration is reduced gradually. This visual representation was created intentionally and its interpretation is that citizens are empowered as they get closer to the highest level of participation. Figure 6 shows that, at the *eEmpowerment* level, the final decisions are now placed in the side of citizens.

Fig. 2 eParticipation
evaluation framework—
eInforming. Adapted from
Terán [4]

Fig. 3 eParticipation
evaluation framework—
eConsulting. Adapted from
Terán [4]

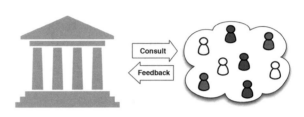

Fig. 4 eParticipation
evaluation framework—
eDiscussion. Adapted from
Terán [4]

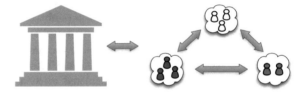

Fig. 5 eParticipation
evaluation Framework—
eParticipation. Adapted from
Terán [4]

Fig. 6 eParticipation
evaluation Framework—
eEmpowerment. Adapted
from Terán [4]

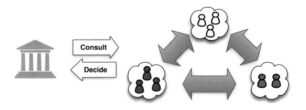

Cognitive cities should include eParticipation when developing new applications
and services for citizens, and to motivate the decision-making process using virtual
channels. In the next section, the architecture of a smart system to enhance participation and provide mechanisms of eEmpowerment with the use of citizen
dynamic profiles.

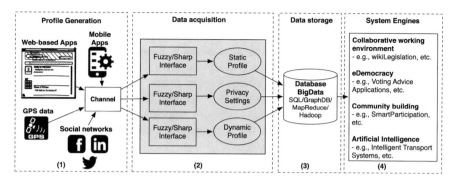

Fig. 7 System architecture for eEmpowerment

3   System Architecture

The system architecture encompasses the stages of eCollaboration, eDemocracy, and eCommunities through which eEmpowerment can be achieved (Fig. 7). It is composed of the following blocks:

- **Profile Generation**. The system proposed is using so-called dynamic profiles, allowing users to become content generators as proposed in the work of Terán [4]. The diverse channels allow citizens to register in web or mobile applications, generate content, participate in forums, and improve citizens' participation. The dynamic profile uses different modules, including privacy and trust definitions, as well as context-awareness, and sentiment analysis, among others.
- **Data acquisition**. Through different types of interfaces (e.g., fuzzy, sharp) data generated from citizens are acquired and stored in the database as static and dynamic profiles, including privacy settings. See block (2) in Fig. 7.

  - **Static Profiles**. These profiles are mainly used to collect personal information from users, such as date of birth, place, education, etc. Static information is required when users subscribe to the system. Users could also include details regarding their preferences, topics and groups of interests in which they would like to participate. Static profiles are predefined by system administrators, so it is a disadvantage that it is very difficult to add additional information if they are modified or extended. A solution to tackle this problem is presented in the work of Alt et al. [10].
  - **Dynamic Profiles**. The collection of dynamic profiles has the goal of improving users' profile generation. Unlike static profiles, the system proposed in this chapter allows users to become content generators. The dynamic profile is created on the basis of the activity, as well as the different content types created via a web interface or mobile applications designed by city administrators. This type of profile is defined to include an additional source of information from users.

- **Data storage**. The system makes use of a relational database to store citizens' profiles, as well as a graph database to store relations and interactions generated by users. To process the dynamic part of the citizens' profiles, big data technologies such as MapReduce for Hadoop are used. See block (3) in Fig. 7.
- **Systems engine**. The system includes a set of hybrid information technologies, particularly a collaborative working environment for eCollaboration, recommender systems, and voting advice applications for eDemocracy, community-building processes for eCommunities, and artificial intelligence, among others. See block (4) in Fig. 7.

4   Case Studies

In this section, three case studies that include the applications of intelligent agents are presented. They intend to highlight how eParticipation of citizens can be increased. The first case explains how collaborative working environments are used in the public sector for a collaborative legislation process, to support people in a virtual space in a time- and place-independent manner in the creation of an organic law. The second application illustrates how eDemocracy processes can be facilitated with the help of a so-called social voting advice application, which takes into consideration trust values among citizens and governmental representatives. Finally, the third application highlights how civic participation can be enhanced through building communities of interest and communities of practice within the *SmartParticipation* platform.

4.1   eCollaboration—A Case Study of Wiki Legislation

Electronic collaboration is a key factor for the success of cognitive city projects proposed by public administration. It has being used in the private sector with the application of so-called collaborative working environments (CWE) to support people's collaboration in a virtual space in a time- and place- independent fashion [2]. The main services that a CWE should include are e-mail, instant messaging, application sharing, video conferencing, collaborative workspace, document management and version control system, task and workflow management, Wiki group or community, and blogging, among others.

Open innovation and open data, urban lags, open sensor networks, and crowdsourcing, among others, together with citizens' engagement, describe some types of collaborative work in Smart Cities. The project *Open Cities*[1] is an example of a collaborative project. It is co-founded by the European Union that aims to

---

[1]http://opencities.net.

Fig. 8 WikiCOESC + i [12]

"validate how to approach Open & User Driven Innovation methodologies to the Public Sector in a scenario of Future Internet Services for cognitive cities" and is running in seven major European cities: Helsinki, Berlin, Amsterdam, Paris, Rome, Barcelona, and Bologna.

As mentioned in Sect. 2, it is important that collaborative environments used by public administration include a level of eEmpowerment, so that citizens are more engaged and feel empowered. In the next section, a case study about a participatory Wiki legislation that was introduced by public administration in Ecuador is described.

4.1.1 Wiki Legislation

In February 2014, the Secretariat of High Education, Science, Technology, and Innovation of Ecuador (SENESCYT [11]) introduced a Wiki-based platform for the creation and discussion of the Organic Code of Social Economy of Knowledge, Creativity and Innovation "Ingenios" [12] that was presented for approval at the National Assembly of Ecuador[2] in May 2015 [13]. The Wiki platform is shown in Fig. 8.

According to SENESCYT, the project proposes a new model of intellectual property for the benefit of the development of the country to promote research and innovation. This initiative of participatory construction of public policy is based on the use of information technologies through citizen participation into

---

[2]http://www.asambleanacional.gob.ec.

Table 2 Most viewed pages. Adapted from Terán et al. [24]

Web pages	Number of visits
Organic code of social economy of knowledge, creativity, and innovation	1,186,668
Discussion: Book III: Knowledge management	125,821
Discussion: Organic code of social economy of knowledge and innovation	70,030
Common provisions	30,907
Discussion: Book II: Social responsible research and innovation	20,774
Book III: Knowledge management	14,342
Book II: Responsible research and social innovation	11,444
Book I: National system of science, technology, innovation and ancient knowledge	10,212

decision-making. For the design of the platform, an open-source Wiki framework was used[3] based on the following criteria: licensing, development of open source, use of open-access tools, easy to use, thematic UI, and interaction with users, among others. The use of the Wiki platform was socialized, prior to making it available online, with different sectors of the society, such as research institutes, universities, artists, authors and composers, innovators, and the production sector, among others. The feedback produced in the socialization was included in the design of the Wiki platform.

Citizens' contributions were collected in section "Discussion" for each item that was created. In order to manage all contributions, the role of reviewer was created. This role included a multidisciplinary team incorporating all contributions with legal correctness to be considered as appropriate. One of the disadvantages presented in this project was the difficulty to determine the number of versions. However, it is estimated that eight versions were made since the platform was launched.

The Wiki legislation project proposed by SENESCYT had over 2 million visitors, 16,391 registered users, more than 1,800,000 visits and exceeded 38,000 editions [12]. Table 2 summarizes the most viewed pages of the Wiki tool.

The project *Ingenios* was conceptually proposed to involve citizens in the creation of laws and to open communication and discussion channels using information and communication technologies via a Wiki platform to accomplish these goals. Wiki tools are designed to enable different users to edit entries quickly and easily in contexts such as collaborative content. Nevertheless, they are not suitable to fulfill all requirements from the public administration, as well as from the side of citizens in a collaborative legislation.

The inclusion of intelligent agents, such as recommender systems, community building, and profile definitions must be included to enhance user's participation. Recommender systems provide citizens with relevant information about topics,

---

[3]MediaWiki platform: https://www.mediawiki.org/.

users, and projects that could be interesting. In this way, citizens are engaged to collaborate more actively. Community-building mechanisms can be used by citizens to create professional or interest-based communities. An example of the use of community building with recommender systems in a collaborative working environment as presented in Sect. 2. The final goal of this approach is that citizens can reach the level of eEmpowerment, which means that the final decision of a project proposed by the public administration could be placed in the side of the society.

To tackle some of the problems described in this section, on August 28, 2015, the Secretary for Higher Education, Science, Technology and Innovation of Ecuador introduced a new project called *Participa* (which means "participate") [14]. The platform allows citizens to review the basis of the plan proposal, submit comments, ask questions, and make contributions.

4.1.2 Maturity Model Level

According to the maturity model presented in Sect. 2, the case study described in this section involves the following levels: eInforming, eDiscussion, and eParticipation. The eInforming level is achieved via social media channels to inform citizens and invite them to participate in the elaboration of the organic code *Ingenios*. The levels eDiscussion and eParticipation are achieved with the use of the Wiki platform proposed by SENESCYT. Even the Wiki tools are designed to allow different users to edit entries quickly and easily in contexts such as collaborative content. With the help of the Wiki tool, Internet users can create, edit, and link individual entries to a topic or document. Nevertheless, they are not suitable to fulfill all requirements from the public administration, as well as from the side of citizens in a collaborative legislation environment. The final goal of this approach is that citizens can reach the level of eEmpowerment, which means that the final decision of a project proposed by the public administration could be placed in the side of the society providing methods of eVoting that would allow citizens to accept or reject specific proposals such as the one suggested in this case study.

4.2 eDemocracy—A Case Study of Voting Advice Application

In the digital era, eDemocracy became an indispensable part of the information society. Online communication is presented as a major source of political information, communication, and participation. eDemocracy sub-processes, such as eDiscussion, eVoting, eElection, and ePosting, show that the use of electronic information and exchange enhances the citizen's involvement, decision-making processes, and stimulates public discussion [2]. eDemocracy is a constituent part of the cognitive city, where the citizen is a key element for the urban governance through civic participation.

This section introduces a web-based voting advice applications (VAA) as an intelligent agent that provides citizens with voting recommendations. As a part of the eDemocracy processes, smart tools such as VAA have been heavily used by citizens during the election period in European countries [15]. The case study presented is a web-based system that extends functionalities of a Swiss VAA (smartvote.ch) towards *social* VAA. The prototype takes into consideration the trust and reputation among citizens and candidates that further influence the final generation of the voting recommendation. The trust among users of the platform enables the system to prevent the appearance of unreliable and untrustworthy users.

4.2.1 Trust-Aware Voting Advice Application

The Trust-Aware Voting Advice Application prototype is a collaborative filtering-based recommender system. The system solves the following tasks for the citizen:

- *Help a citizen to find other citizens s/he might like.* This helps with forming discussion groups, matchmaking, or connecting users so that they can exchange recommendations socially [16]. In this connection, the prototype provides recommendations of citizens who have the same political attitudes and opinions as a target user, and citizens who have the trustworthiness and the reputation within particular political community.
- *Help a citizen to find political issues that might be interesting or worthy of considering.* The prototype provides recommendations of political issues that might be interesting for the citizen, thus enhancing people's interest with further discussion about these policies in the forum. Providing a recommendation of a new policy brings serendipity and new discoveries for the citizen.
- *Advise a citizen on a particular political candidate.* A citizen has a particular candidate in mind; would the community vote for her/him? In this case, the system supports citizens when making decisions about which candidate to vote for or not. As a result, the system recommends candidates for the citizen based on the community's choice.

The implementation of the prototype is described in detail in the work of Radovanovic [17] and it is available at [18]. It uses a Q2A platform, PHP and JavaScript, while for the database management MySQL is used. The objective of the prototype is to use a trust-aware recommender technique in order to preserve the privacy of the system and to protect it from unreliable users. For this reason, the trust network between users of the platform is built. The trust network computation is based on the global and local trust metrics that use two propagation methods: propagation by average and propagation by multiplication [19]. With inclusion of trust values, the system generates final recommendations. The recommendation

Fig. 9 VAA—eDiscussion forum

output is based on the politically similar citizens, as well as on the trustworthy citizens and the combination of both communities. The following functionalities are implemented:

- **eDiscussion forum** enables citizens' interaction through questions and answers. The eDiscussion channel allows to citizens and candidates to debate, exchange, and share their opinion on particular political topics or issues. Citizens and candidates are able to discuss political topics or issues of their interest by asking a question, and by providing answers and comments on particular questions (see Fig. 9).
- **Voting questionnaire** is used to collect information about citizens' political opinions, upon which the political profile is created. In this module, all available political issues are listed. Citizens can choose the category of questions, related to some political issue that s/he wants to answer. While the activities of each citizen on the discussion forum are visible, the answers of a citizen provided in questionnaire module are not visible to other citizens (see Fig. 10).

Forum   Questionnaire   Recommendation

Questionnaire

Welfare, family and health

- *Do you support the raising of the pensionable age to 67 for men and women?*
  Yes   Rather yes   Rather no ● No   Weight ++ ⬍
- *Should health insurance premiums for basic insurance be calculated according to the ability to pay (based on income and property)?*
  Yes ● Rather yes   Rather no   No   Weight - ⬍

Fig. 10 VAA—voting questionnaire

Fig. 11 VAA—recommendation of citizens

- **The recommendation** section shows the recommendation lists of citizens and community-based political issues. The following types of recommendations are generated by the system:

  – **Recommendation of citizens** is generated by calculating the similarity between citizens and calculating the trustworthiness of citizens. For the similarity computation, a similar approach as in smartvote.ch is adopted [20]. This approach is based on the Euclidean distance. After calculating the Euclidean distance, this value is normed using the value of maximal possible distance, and converted to a percentage. For the trustworthiness value, the citizen is asked to explicitly express her/his level of trust in some other citizens. The global reputation of the citizen within eDiscussion forum is also taken into consideration. The citizen sets the percentage of trust/similarity inclusion in the final recommendation score, after which the lists of similar/trustful citizens are provided (see Fig. 11).

  – **Issues recommendation using average voter approach** calculates the distance between the citizen and the average voter. In the work of Katakis et al. [21], this algorithm concerns the average voter of a party. In this prototype, the average voter of a certain political issue is considered. For each issue, the average voter is calculated, and the Euclidean distance between the citizen and the average voter of that political issue is calculated. Furthermore, average trust per political issue is taken into consideration for the final issue recommendation. The citizen sets the percentage of trust/similarity inclusion in the final recommendation score. After that the list of political issues with the greatest score is provided (see Fig. 12).

  – **Issues recommendation based on a citizen's activity in eDiscussion forum** is based on the number of activities that s/he has in each political category. The number of activities includes the number of questions that a

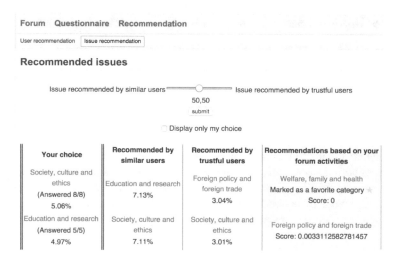

Fig. 12 VAA—recommendation of political issues

citizen asked, number of answers that s/he provided in the discussion forum, the number of comments, votes, and the number of questions that were marked as interesting, then the calculated recommendation is provided as a score list (see Fig. 12).

The main goal of the prototype presented is to enrich VAAs with eDiscussion channel where citizens and candidates are able to exchange and share with their opinion about upcoming or current referendums or political issues. Within eDiscussion activities of citizens and candidates, a trust network is aggregated and used for the trust-enhanced generation of the recommendation. Thus, the risk of the unreliable and untrustworthy citizens that might influence the recommendation process is reduced.

4.2.2 Maturity Model Level

The case study presented involves eInforming and eDiscussion levels of the maturity model discussed in Sect. 2. Particularly, within the Questionnaire part of the prototype, citizens are informed and provided with relevant information about different policies and referendums. The eDiscussion level of the maturity model is achieved via the forum part of the prototype where discussions channels are established between citizens and governmental representatives (candidates). The prototype partially covers eParticipation level, which is represented by citizens' engagement within the platform; however, the collaborative working environment on public projects of interest is missing to entirely complete eParticipation level.

4.3   eCommunities—A Case Study of SmartParticipation

Nowadays, the Internet is developing into an environment in which the citizens are able to display themselves in a virtual world, meet with other citizens, exchange information and services, promote common projects, and overcome language and cultural boundaries with the use of communication and information technologies, thanks to the introduction of the so-called Web 2.0 and the use of social software.

Computers and communication channels are not only used by administrators with technical skills to upload content. Regular users and collaborators are content generators with the possibility to make encounters and communities possible. In the same way, physical locations such as street cafes, markets or exhibitions are considered as points of encounter, cyberspace has developed as a virtual location. Topic-specific, cultural or scientific meeting points on the Internet create a new kind of community formation.

Computer networks are populated by citizens and avatars, in many cases anonymously. The Internet can amplify the citizens' living environment into a virtualized world. For the development of cognitive cities, it is important to provide citizens with virtual spaces where they can express themselves and contribute to the development of the city. Among the communities created on the Internet, we can distinguish between two kinds:

- **Communities of interest**: This includes citizens who are interested in a common thing or share a hobby.
- **Communities of practice**: This includes groups of citizens who participate in a common project of a governmental institution and invest time and knowledge.

Both kinds of communities can be furthered by information and communication systems. Community support systems serve the members as a meeting point, or to exchange know-how and knowledge and to master tasks or challenges. Web-based platforms not only reveal the existence of communities, but also make it possible to meet other members of the community and to utilize the know-how and expertise of this community. Web-based platforms and the corresponding software systems for community formation can be characterized as follows:

- **Civic Network Systems**: Civic networks or community networks are electronic meeting points for citizens whose common ground is a place or living environment they have in common. This may be for example, a city or a mountain area whose inhabitants want to meet with their cohabitants in the virtual space. Apart from designing and moderating discussion forums, the focus lies on projects that concern the community or on improved and extended possibilities of training and further education.
- **Buddy Systems**: Based on the everyday meaning of the word "buddy", the buddy system shows where colleagues or friends are currently located and how they can be reached electronically. The social or task-oriented perception of group members (awareness) makes it possible to meet virtually or exchange experiences; apart from that, the system indicates it if a participant does not want

to be disturbed. For actual reunions, the systems can establish audio or video connections so that the spatial distance is overcome (media space). Recent developments admit to creating three-dimensional worlds and to replace real areas of encounter with virtual ones (virtual reality).

- **Matchmaking Systems**: The term "matchmaking" originally refers to marriage bureaus, but is conceived as a wider concept in cyberspace. It is about exchanging relations for economy and society. The systems promote contacts and activities in a commonly used environment for proper purposes. For example, networks of acquaintances are utilized to make new contacts, based on an already existing mutual trust, and to exchange information.
- **Recommender Systems**: These systems are about finding out the Internet users' preferences and making them suggestions for their further development. Special procedures (collaborative filtering) make it possible to categorize the participants' preferences and to pass on those suggestions for activities and further education that are important for a certain group. For example, if someone wishes to become acquainted with a new subject area, such systems recommend suitable literature, possibly complemented by an expert evaluation.
- **Corporate Blog Systems**: Corporate Weblogs are up-to-date digital journals of groups of people or organizational units. The systems or platforms make it possible to mediate knowledge, occupy topics, or nurture relationships. Corporate blogs serve to support the organizational goals and usually can be subscribed to by all interest groups.

4.3.1 SmartParticipation

The prototype presented in this section on *SmartParticipation* is described in detail in the work of Eifert [22] and is available at [23]. It uses a graph database as the kernel for the design and implementation of a recommender system and allows users to receive real-time recommendations. Key points of the implementation are the use of dynamic profiles to enable users to become content generators. Dynamic profiles are updated automatically, always reflecting the current state of the user and displaying recommendations through collaborative filtering. Three types of recommendation are implemented: popular articles, recommended articles, and users. The prototype presented in this section includes the following five elements:

- *Web interface*. It is the front end of the system proposed and allows users to register, generate content, comments, ratings, and interaction, among others (refer to Fig. 13).
- *Data acquisition*. It includes different types of interfaces (e.g., sliders, stars, binary, etc.) for rating the content generated by users. These ratings are stored in the different databases and used for the generation of recommendations (refer to Fig. 14).

Fig. 13 *SmartParticipation*—web interface

Fig. 14 *SmartParticipation*—data acquisition

- *Data storage.* The system developed makes use of a relational database to store ratings, as well as graph databases to store relations and interactions generated by users.
- *Recommender engine.* It includes a set of hybrid recommendation system methods, such as collaborative filtering and a fuzzy-based recommender [4], using similarity metric that is used to provide recommendations to users.

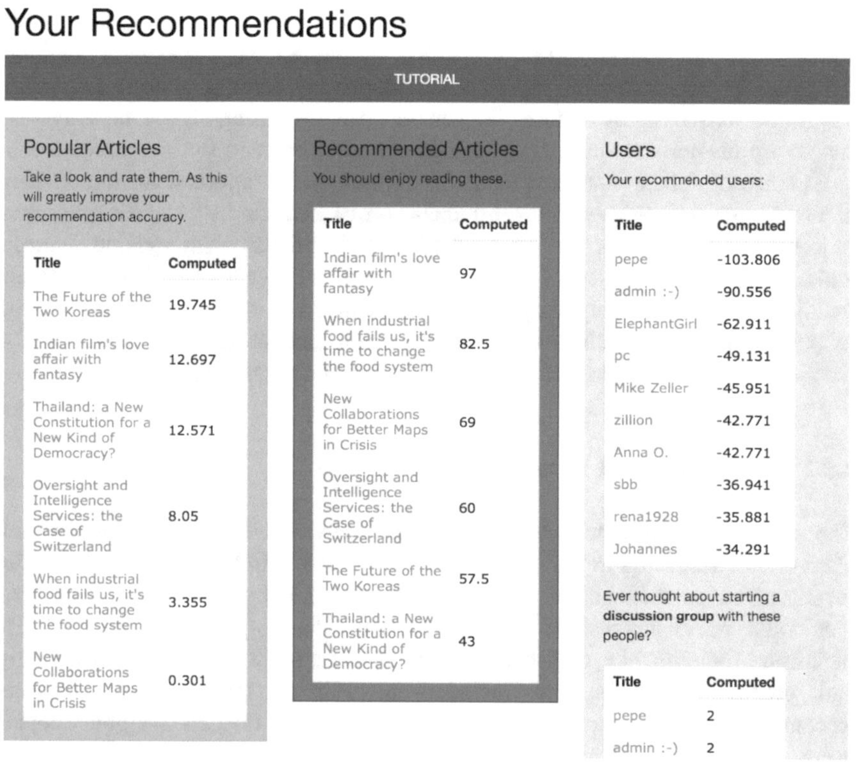

Your Recommendations

TUTORIAL

Popular Articles

Take a look and rate them. As this will greatly improve your recommendation accuracy.

Title	Computed
The Future of the Two Koreas	19.745
Indian film's love affair with fantasy	12.697
Thailand: a New Constitution for a New Kind of Democracy?	12.571
Oversight and Intelligence Services: the Case of Switzerland	8.05
When industrial food fails us, it's time to change the food system	3.355
New Collaborations for Better Maps in Crisis	0.301

Recommended Articles

You should enjoy reading these.

Title	Computed
Indian film's love affair with fantasy	97
When industrial food fails us, it's time to change the food system	82.5
New Collaborations for Better Maps in Crisis	69
Oversight and Intelligence Services: the Case of Switzerland	60
The Future of the Two Koreas	57.5
Thailand: a New Constitution for a New Kind of Democracy?	43

Users

Your recommended users:

Title	Computed
pepe	-103.806
admin :-)	-90.556
ElephantGirl	-62.911
pc	-49.131
Mike Zeller	-45.951
zillion	-42.771
Anna O.	-42.771
sbb	-36.941
rena1928	-35.881
Johannes	-34.291

Ever thought about starting a **discussion group** with these people?

Title	Computed
pepe	2
admin :-)	2

Fig. 15 *SmartParticipation*—recommendation output

- *Recommendation output.* The system provides different types of outputs of recommendations, as well as visualizations. Users can then get recommendations on the following items (refer to Fig. 15):

  - Users with whom he shares a common opinion. This allows users to create activist groups or political parties.
  - Users with whom he has had lots of interaction. This allows users to create a discussion group with these individuals.
  - A combination of both items (users) above, based upon a customization setting of the recommendation seeker.
  - Articles which the user finds interesting and will probably rate highly.
  - What he should be rating in order to get more accurate recommendations. This is especially important for new users or those who have been inactive for some time.

Popular articles are non-personalized recommendations that help users to identify popular content. Recommended articles and users are based on interaction and ratings of different posts. The system allows users to customize their recommendation using interaction- or opinion-based settings. The system allows active users to contact other users for discussion or community building. Additionally, users are able to set preferred privacy settings and are allowed to contact an active user by defining his personal privacy settings. Thus, an active user can set up his personal profile as open or closed to be used for recommendations.

The use of debate platforms *SmartParticipation* are important for the development of cognitive cities seeking for inclusion of citizens in the decision-making process. Such platforms can be used to engage citizens with specific interests, knowledge, and expertise. The use of recommender systems and community building processes are important tools to be included on debate platforms, in order to potentially provide electronic voting mechanisms or look for a decision or feedback coming from expert communities on a statement or proposal.

4.3.2 Maturity Model Level

The case study presented in this section involves the levels eInforming and eDiscussion according to the maturity model presented in Sect. 2. The eInforming level is achieved via social media channels to inform citizens about different issues that could be of interest of different citizens according to their profiles. The inclusion of intelligent agents, such as recommender systems, community building, and profile definitions, are included to enhance participation of users. Recommender systems provide citizens with relevant information about topics, users, and projects that could be interesting. In this way, citizens are engaged to collaborate more actively. Community-building mechanisms can be used by citizens to create communities of interests or professional communities. Nevertheless, other levels of the maturity model can be included, such as eParticipation and eEmpowerment to engage citizens to better participate in projects and to include mechanisms for eVoting to place the decision-making on the side of citizens.

5   Conclusions

Cognitive Cities are defined as those that use information and communication technologies to provide services in urban cities to enhance the citizens' quality of life and also to optimize the costs and resources that are used to provide services to citizens. Cognitive city projects are developing faster in areas such as eMobility, energy management, health, intelligent buildings, innovation labs, big data, safety, security, and privacy, among others.

In this chapter, the authors used an eGovernment framework that includes three levels, including information and communication, production, and participation. All services mentioned above can be located in the first and second levels of this framework, more specifically to the level of production where all services provided to citizens are located. The third level of participation provides mechanisms to enhance the involvement of citizens with the use of intelligent agents such as recommender systems, community-building processes and electronic voting, in order to give citizens the opportunity to decide on relevant issues and projects that could be proposed from the administration as well as the society.

Three projects are presented to support the concepts provided in this chapter. The first one corresponds to a Wiki Legislation developed by the Secretariat of Higher Education, Science, Technology and Innovation of Ecuador [11]. The so-called *Ingenios Code* proposal was built to aim at the democratization of participation and access to knowledge. The second project is an extension of voting advice applications in a dynamic context, including trust values among citizens and governmental representatives. The third project makes use of recommender systems for community-building processes to provide communication channels and a debate platform.

Cognitive cities should consider the creation of new services for citizens by including them in decision-making processes via information and communication technologies toward the so-called eEmpowerment. Thus, administration can profit from the knowledge of the crowd, and citizens on the other hand would be engaged to participate in different initiatives coming from the society, administration, and private sector.

Acknowledgements The authors would like to thank the members of the Information System Research Group at the University of Fribourg (http://diuf.unifr.ch/is) for contributing valuable thoughts and comments. Special thanks to the office of the Secretariat of Higher Education, Science, Technology and Innovation of Ecuador for its support and collaboration with relevant information on the case study *Wiki Legislation* presented in this chapter.

References

1. Mostashari, A., Mansouri, M., Arnold, F., Finger, M.: Cognitive cities and intelligent urban governance. Netw. Ind. Q. **13**(3) (2011)
2. Meier, A.: eDemocracy and eGovernment: stages of a democratic knowledge society. Springer-Verlag, Berlin Heidelberg (2012)
3. Schaffers, H., Komninos, N., Pallot, M., Trousse, B., Nilsson, M., Oliveira, A.: Smart cities and the future internet: towards cooperation frameworks for open innovation. Future Internet Assembly **6656**, 431–446 (2011)

4. Terán, L.: SmartParticipation: a fuzzy-based recommender system for political community-building. In Fuzzy Management Methods, Springer (2014)
5. Tambouris, E., Liotas, N., Tarabanis, K.: A framework for assessing eParticipation projects and tools. In: Proceedings of the 40th Annual Hawaii International Conference on System Sciences, HICSS '07, p 90. IEEE Computer Society, Washington, DC, USA (2007). doi:10.1109/HICSS.2007.13
6. Palen, L., Liu, S.B.: Citizen communications in crisis: anticipating a future of ict-supported public participation. In: Proceedings of the SIGCHI Conference on Human Factors in Computing Systems, pp. 727–736. ACM (2007)
7. Ae Chun, S., Luna-Reyes, L.F., Sandoval-Almazán, R., Ae Chun, S., Luna-Reyes, L.F., Sandoval-Almazán, R.: Collaborative e-government. Transforming Gov. People, Process and Policy 6(1), 5–12 (2012)
8. Wang, Y., Ruhe, G.: The cognitive process of decision making. Int. J. Cogn Inf. Nat. Intell. (2007)
9. de Souza, G.P., Maciel, C.: The voting processes in digital participative budget: a case study. In: 3rd International Conference on Electronic Voting 2008, Gesellschaft für Informatik (GI), pp. 6–9 (2008)
10. Alt, F., Balz, M., Kristes, S., Shirazi, A.S., Mennenöh, J., Schmidt, A., Schröder, H., Goedicke, M.: Adaptive user profiles in pervasive advertising environments. Ambient Intelligence. Lecture Notes in Computer Science, vol. 5859, pp. 276–286. Springer, Berlin Heidelberg (2009)
11. SENESCYT.: Secretara de Educación Superior, Ciencia, Technologa e Innovación. (Online) Available at http://www.educacionsuperior.gob.ec (2015). Accessed 15 Sept 2015
12. SENESCYT.: Código orgánico de economa social del conocimiento e innovación. (Online) Available at http://coesc.educacionsuperior.gob.ec (2015). Accessed 15 Sept 2015
13. IFTH: Código ingenios fue presentado oficialmente, previo a su trámite en la asamblea nacional. (Online) Available at http://www.fomentoacademico.gob.ec/codigo-ingenios-fue-presentado-oficialmente-previo-a-su-tramite-en-la-asamblea-nacional/ (2015). Accessed 15 Sept 2015
14. SENESCYT.: Participa. mi ecuador, nuestro futuro. (Online) Available at http://participa.ec/ (2015). Accessed 09 Oct 2015
15. Garzia, D., Marschall, S.: Voting advice applications under review: the state of research. Int. J. Electron. Gov. 5(3), 203–222 (2012)
16. Schafer, J.B., Herlocker, J., Frankowski, D., Sen, S: Collaborative filtering recommender systems. In: The Adaptive Web, pp. 291–324. Springer, Berlin (2007)
17. Radovanovic, N.: Integration of trust-awareness in political recommender systems. Master's thesis, Information Systems Research Group, University of Fribourg, Switzerland (2015)
18. Radovanovic, N.: TrustVAA. (Online) Available at http://trustrs.isproject.ch/ (2015). Accessed 15 Sept 2015
19. Ricci, F., Rokach, L., Shapira, B., Kantor, P.B. (eds): Recommender systems handbook. Springer, US (2011)
20. Smartvote.: Voting advice application. (Online) Available at http://smartvote.ch (2003). Accessed 22 June 2015
21. Katakis, I., Tsapatsoulis, N., Mendez, F., Triga, V., Djouvas, C.: Social voting advice applications-definitions, challenges, datasets and evaluation. IEEE Trans. Cybernetics 44(7), 1039–1052 (2014)
22. Eifert, J.: Smartparticipation—generating dynamic profiles based on user interaction with the help of a graph database. Master's thesis, Information Systems Research Group, University of Fribourg, Switzerland (2015)

23. Eifert, J.: SmartParticipation. (Online) Available at http://smartparticipation.herokuapp.com (2015). Accessed 15 Oct 2015
24. Terán, L., Spicher, N., Ramrez, R., Pazos, R., Ron, M.: Public collaborative legislation. A case study of the Ingenios Act. In: International Conference on eDemocracy & eGovernment, ICEDEG 2016, IEEE Computer Society (2016)

Cognitive Cities, Big Data and Citizen Participation: The Essentials of Privacy and Security

Ann Cavoukian and Michelle Chibba

Abstract Our message for using humans as sensors to enable a Cognitive City is the same as that for any emerging technology: privacy cannot be sacrificed for other anticipated benefits. Privacy protections will be critical to the adoption of Cognitive City sensor technologies—individuals must feel comfortable that their privacy will not be violated as they move about in public spaces. We do not propose stifling innovation or denying the value of technology. We urge adopting a positive-sum paradigm: allowing both privacy *and* functionality to co-exist. Adopting privacy early at the design phase will achieve this goal. We advance the view that Privacy by Design is the sine qua non for all advances in technology, data management and application of cognitive technologies, as envisioned by Cognitive Cities. The 7 Foundational Principles of Privacy by Design can pave the way to building the necessary trust and accountability required in the systems involved.

Keywords Privacy by design · Privacy · Data protection · Security · Big data · Surveillance · Cognitive computing · SmartData

1 Introduction

Over the last two decades, the world has experienced rapid urbanization and at this rate of growth, the predictions are that roughly three-quarters of the world's population will soon be living in cities. Considerable thought is being given to how cities will deal with sustainability challenges associated with such growth. Across the globe, we are witnessing cities that are already implementing strategies to modernize their critical infrastructure and services in order to meet the growing citizen demands. Central to this vision of sustainability and modernization is the use

A. Cavoukian (✉) · M. Chibba
Privacy and Big Data Institute, Ryerson University, Toronto, Canada
e-mail: ann.cavoukian@ryerson.ca

M. Chibba
e-mail: michelle.chibba@ryerson.ca

© Springer International Publishing Switzerland 2016
E. Portmann and M. Finger (eds.), *Towards Cognitive Cities*,
Studies in Systems, Decision and Control 63, DOI 10.1007/978-3-319-33798-2_4

of networked infrastructures and information communication technologies (ICTs) which are the underpinnings of a "Smart City."

With the advent of these technologies, however, the rules for managing and interacting with data [1] have changed radically. The global creation of data is accelerating, leading to discussions on how to unlock the value of Big Data. Indeed, the "Internet of Things (IoT)" is quickly being overshadowed by the 'Internet of Everything.' Data mobility forces us to focus on data in motion, not just data at rest or in use. These trends carry profound implications not only for "Smart Cities" but also for privacy, data protection and data security.

It is apparent that there is a growing interest [2] in moving beyond variants of technology-based information-centric Smart Cities to an infrastructure that includes 'cognition' or "elements of learning, memory creation and experience retrieval for continuously improving urban governance." Our view is that "design-thinking" should influence innovation, creativity and productivity. In the context of Cognitive Cities, privacy must be approached from the same design-thinking perspective. Privacy must be designed into every technology, system, standard, protocol and process that touches the lives and identities of citizens in a Cognitive City.

This chapter seeks to make possible the coexistence of privacy and cognitive data driven technologies—the cornerstone of a Cognitive City, through a universal framework for the strongest protection of privacy available in the modern era, known as Privacy by Design (PbD). First, we make the case for privacy as an eternal human value and therefore integral to any discussion of networked infrastructure, ICTs, the Internet of Everything and Big Data. Second, we discuss Privacy by Design as a framework for modernization, specifically in the context of a Cognitive City, and lastly, provide summaries of three relevant examples of systems and technologies that have incorporated Privacy by Design.

2 The Cognitive City: A Case for Privacy as an Eternal Value

In a 2013 speech given by Sennett [3] he states, "The cities everyone wants to live in should be clean and sage, possess efficient public services, be supported by a dynamic economy, provide cultural stimulation, and also do their best to heal society's divisions of race, class, and ethnicity. These are not the cities we live in." Indeed, major cities around the world are grappling with unprecedented growth, ageing infrastructures, traffic congestion, pollution, affordable housing shortages, to name a few of the ecological, economic and social pressures. The facts on urbanization are staggering—a UN report on world urbanization [4] states that the urban population has increased from 746 million in 1950 to 3.9 billion in 2014 and projects a further 2.5 billion more people by 2050.

In 1999, almost two decades ago, a National Academy of Engineering (NAE) newsletter focused on the engineering challenges facing American cities, using New York and Boston as examples and how science, technology and engineering may contribute to enhancing the quality of city life. In one of the NAE newsletter articles entitled "Wild Ideas for Future Cities," [5] Joseph F. Coates pointed to smart homes and cars as examples to demonstrate how "smartness could radically alter physical infrastructure."

Fast forward 15 years where existing cities are revamping their critical infrastructure and services through advances in technology (e.g. Chicago; Toronto; the Hague), and new cities are being built from scratch to be 'smart' (e.g. Masdar City, Abu Dhabi; SmartCity Malta). ICTs such as RFID, Wi-FI, ZigBee, Machine-2-Machine communication, asset management, smart meters, sensor networks, [6] for example, are some of the technological components being deployed.

Now imagine a city that is able to learn, sense and adapt to different experiences and ecosystem changes [2, 7] where single processes are made more cognitive and collectively make up a framework of intelligent urban governance. This is the vision of a "Cognitive City." A Cognitive City is said to go beyond the vision of a Smart City because it enables involvement, generates awareness, uses creativity, deals with uncertainty, and takes a holistic approach to involving its citizens. The advantage of taking such an interdisciplinary approach [8] is its ability to deal with complex and diverse urban issues like space, economy, technology, culture, politics as well as religion, education and behavior.

As the ecosystem from 'smart' to 'cognitive' evolves, this will allow for the reshaping of the urban governance landscape [9] and how services are produced. The notion of "citizens as human sensor networks" [2] by way of our ubiquitous attachment to our smartphones is indeed far reaching. Mostashari et al. envision a city where "the citizen becomes an active element of urban governance, not only through civic participation, but also through serving as a sensor for the operational state of the urban infrastructure." Information technology and embedded intelligence through cognitive computing and ambient intelligent systems are considered to be the necessary ingredients [9, 10] that will allow citizens and government to drive innovative service provision.

For citizens to be active players and active co-producers of services, they need to be much more informed, aware and involved [11]. This requires governments to move away from traditional information management where information is highly compartmentalized or guarded, to greater openness and sharing of data. This shift is not only dramatic but also raises concerns [11] related to privacy and data security. Mostashari et al. acknowledge that "An important consideration for leveraging citizens as information providers in the urban environment is the issue of data privacy and security." The success of the Cognitive City rests on data and information sharing by active citizens [2] and "if only 1–2 % of the urban population is willing to play an active role in the cognitive grid in exchange for better information access on urban infrastructure services, the implications would be dramatic."

This section provides a more detailed discussion of the importance of privacy in the larger socio-political sphere and as an eternal value necessary for the success of a Cognitive City. It will also include a review of the major privacy and data security issues associated with a networked infrastructure and ICTs that are integrally involved in achieving a Cognitive City. These are issues raised among regulators, academics, businesses, governments, media and the general public arising from Big Data, the Internet of Everything, cloud computing, sensor technologies, social media, Web 2.0, body worn technologies (wearables), other emerging technology trends and advancements in cognitive computing. Identifying such risks and recommending potential mitigation strategies will be the necessary first steps in any discussion of the technologies that will support the vision of a Cognitive City.

2.1   The Dimensions of Privacy and Significance to Freedom

Privacy has multiple dimensions. Clarke [12] elaborates on at least four dimensions: (i) privacy of the person or the control over decisions affecting one's body such as decisions about blood transfusions without consent; (ii) privacy of personal behavior such as control over one's religious practices whether in private or public; (iii) privacy of personal communications or the expectation that our communications will be under our control and not routinely monitored and; (iv) privacy of personal data or informational privacy.

At the core of informational privacy is the definition of personal information or personally identifiable information, which excludes properly de-identified [13] and aggregated data. The terms "informational privacy" and "data protection" refer to differing but closely related concepts. Privacy is a much broader concept than data protection. Information privacy refers to the right or ability of individuals to exercise control over the collection, use and disclosure by others of their personal information. Data protection is generally established through a set of rules or legal frameworks that impose responsibilities on organizations that collect, use, and disclose personal information. It is important to understand the distinction between the two functions. The first approaches privacy from the perspective of the individual data subject, the second from the perspective of the custodial organization.

Another dimension is that of privacy of location and space. This involves individuals having the right to move about in public or semi-public spaces without fear of being identified, tracked or monitored. This dimension of privacy also includes a right to solitude and a right to privacy in private spaces such as one's home or one's car. All of these dimensions of privacy have social value and relevance for the Cognitive City. When citizens are free to move about in public spaces [14] without fear of identification, monitoring or tracking, they experience a sense of living in a democracy—a sense of living in freedom.

Not only are there many dimensions to privacy but it is also dynamic and fluid in nature, often influenced by context—situational factors, social norms, environmental conditions, etc. Context is unique and highly connected to the individual—it will vary from one person to another. According to Helen Nissenbaum's theory of contextual integrity [15], context is "who is gathering the information, who is analyzing it, who is disseminating it and to whom, the nature of the information, the relationships among the various parties, and even larger institutional and social circumstances." Nissenbaum posits two types of informational norms to which systems should adhere: norms of appropriateness (refers to what an individual is willing to disclose in a particular context such as a patient disclosing highly sensitive information to a physician) and norms of flow or distribution (the movement and or transfer of information from one party to others).

More broadly, privacy underpins freedom. Privacy relates to freedom of choice and exercising control in the sphere of one's identity or self—making choices regarding what personal information one wishes to share and, perhaps more importantly, what information one does *not* wish to share with others. It is this freedom of expression and freedom of association that forms the basis of an open and democratic society. The opposite of freedom [16, 17] would be forcing law-abiding citizens into social conformity and limiting behaviors, knowing that they were being watched.

Privacy [18] also preserves an "essential space for the development of ethically grounded citizens capable of engaging in the critical functions of public citizenship." In a 2012 case discussing the right to "public privacy," the Ontario Court of Appeal [19] stated that "personal privacy protects an individual's ability to function on a day-to-day basis within society while enjoying a degree of anonymity that is essential to the individual's personal growth and the flourishing of an open and democratic society."

Scholars [20] view privacy as "a cornerstone of contemporary Western society because it affects individual self-determination; the autonomy of relationships; behavioural independence; existential choices and the development of one's self; spiritual peace of mind and the ability to resist power and behavioural manipulation." Indeed, in the information and technology era we live in, the protection of our right to informational privacy is increasingly critical to the preservation of our rights to life, liberty, and security of the person—in essence, to preserving our freedom.

2.2 Big Data, Ubiquitous Computing and the Value of Personal Information

Without a doubt, we are living in a world where untold amounts of data are being produced, thanks to networked infrastructures and ICTs not to mention ubiquitous computing and ambient intelligent systems. There are currently 9.6 billion Internet-connected devices, 1.3 billion mobile broadband connections, and 1.2 zettabytes of annual global IP traffic. Every two days, our use of these devices creates roughly five exabytes of data—as much as all the data created by humans from the dawn of civilization to 2003. The result is what has now become known as the data revolution or the era of "Big Data."

While much of this data has no relationship to any given individual, the magnitude of personal data created is unparalleled. While it is common to think of personal information as basic tombstone data (e.g. name, address, phone number), the digitization of data has caused the definition of personal information to expand. It now includes, for example, biological, genealogical, historical, transactional, locational, relational, computational, vocational, or reputational information. Grey areas are also arising from the collection of metadata. In the case of our internet communications, the detailed pattern of associations revealed through metadata [21] can be far more revealing and invasive of privacy than merely accessing the content of one's communications.

At one time, most of our personal data was passively being collected, used and disclosed by organizations. Now, through social media and other emerging Web 2.0 technologies, we are more active producers and sharers of our data. This phenomenon has given rise to a new science of human data interaction (HDI). Indeed, researchers Mortier et al. [1] believe that HDI will supersede the long-standing discipline of human-computer interaction (HCI) which focuses on our interactions with computers such as usability features, user-interface design elements and the like.

In 2009, then European Consumer Commissioner, Kuneva [22] coined an important phrase "Personal data is the new oil of the Internet and the new currency of the digital world." She was speaking of the boundless opportunities that this new treasure trove of data would bring, both economically and socially. The idea of democratizing data has only been made possible because of information technology and the Internet. The desire to unlock the value of personal data has been the subject of a number of global industry initiatives, one of which was initiated by the World Economic Forum [23] as early as 2010, culminating in a series of reports and strategies.

As a consequence, cognitive computing is fast becoming an emerging area that will act as a game-changer not just in unlocking the value of the tsunami of data but transforming the way we interact with technology. Cognitive computing systems [24–26] may be enhancements for the user, or they may act virtually autonomously in problem-solving situations that extend beyond simple data-mining techniques. The value of these systems to enterprises is that it allows them to capitalize on vast amounts of data, untapped because of the limited capacity of human review and to use the derived knowledge and intelligence to challenge conventional ways of doing business.

2.3 The Privacy Imperative

Although this backdrop of the digital revolution forms the basis for optimism in the face of the growing needs and strains facing cities, these same trends and factors are causing a crisis for informational privacy and data protection. The crisis [27] stems from the fact that individuals, with the growth of networked infrastructures and ICTs, no longer have complete control over one's own personal information. The potential exists for technology to become a surveillance tool that will diminish individual privacy, dignity and freedom. The miniaturization of devices used for data collection and the sophistication of sharing technologies are no longer in the realm of science fiction. Concerns about online privacy arise from statements such as the one made by Google Chairman, Eric Schmidt in an interview with the Atlantic (2010): "Google policy is to get right up to the creepy line and not cross it…. With your permission you give us more information about you, about your friends, and we can improve the quality of our search. We don't need you to type at all. We know where you are. We know where you've been. We can more or less know what you're thinking about."[1]

Recent research into HCI and building context-aware applications for mobile devices [28] found that despite solutions to address the difficulties in developing context-aware applications, usability concerns needed to be addressed. Users are concerned about lack of control, lack of transparency and more importantly privacy —concern about how their location and behavior context data may be used by third parties. So despite the promise of these technologies, in the context of a Cognitive City, there could be a backlash by citizens if their privacy is increasingly invaded, thereby diminishing any positive gains or benefits to be achieved.

As data mobility increases vertically and horizontally, there is also less transparency for the individual to make informed decisions about the uses of their data. By removing the individual to whom the data relates [29], the potential for questionable data quality increases, as do false positives, lack of causality, inference-dependency and greater bias in the results. With technologies and systems receding so far into the background, researchers Bellotti and Sellen [30] speak to the challenges of "disembodiment from the context and dissociation from one's actions" where users of communication technologies either forget, get confused or become so comfortable with the technology and associated implications so as to interfere with conveying information about oneself or gaining information about others. The example used from their research project for disembodiment is during an audio-visual communication, without any feedback on what the technology is communicating, the individual may be unaware of the extent of what is being communicated about themselves.

Asymmetries of knowledge tend to foster asymmetries of power manifested by questionable data quality, lack of causality, inference-dependency, bias and false positives. Armed with greater and more detailed knowledge about its citizens, government organizations [11] can embark on social engineering and manipulation, at an unprecedented scale. This is commonly referred to as 'function creep' or secondary use where personal data is collected for one purpose but used for another, without the knowledge or consent of the individual. Pekarek et al. [31] coined the term "surveillance as a service" for this phenomenon. This is a practice that can be described as the proactive use of collected surveillance data, which is enabled by

---

[1]http://www.theatlantic.com/technology/archive/2010/10/googles-ceo-the-laws-are-written-by-lobbyists/63908/.

recent developments in technology and data matching practices that blurs the line between enforcement and administrative authorities of government.

The example given by Pekarek et al. is when the Dutch Tax Authority used Automated License Plate Recognition data to identify taxi drivers who may have exceeded the threshold for allowable kilometres in order to qualify for a tax exemption. The Dutch Tax Authority's use of this law enforcement data was administrative in nature because it sent out notices ahead of the tax filing deadline, to these drivers, alerting them to check that they were still compliant with this requirement. This mechanism implies that license plate surveillance data, normally employed ex post as evidence against suspected offenders, is now used ex ante and proactively to 'remind' non-suspects to be law-abiding citizens for tax purposes. This may lead to the assumption that the service is actually an element of an encompassing surveillance and enforcement strategy. Even if this proves not to be the case, the nature and origin of the data make the service problematic because data are used in a context different from the one in which they were originally gathered.

The volume of digitized data held in databases by organizations as well as publicly available data challenges privacy in several ways. The urge is to move towards maximizing data collection, storage and retention rather than following a data minimization approach central in privacy principles and laws around the world. As Zittrain [32], a renowned Harvard law professor and digital freedom advocate noted "....Government databases remain of particular concern, because of the unique strength and power of the state to amass information and use it for life-altering purposes. The day-to-day workings of the government rely on numerous databases, including those used for the calculation and provision of government benefits, decisions about law enforcement, and inclusion in various licensing regimes."

Other privacy legal scholars and advocates, for example, Solove [33, 34] resurrect the theme in Kafka's "The Trial" to illustrate the prospect of unwarranted government intrusion in a totalitarian regime. Solove expounds on the metaphor of The Trial noting "The problems captured by the Kafka metaphor… are problems of information processing–the storage, use, or analysis of data–rather than information collection. They affect the power relationships between people and the institutions of the modern state. They not only frustrate the individual by creating a sense of helplessness and powerlessness, but they also affect social structure by altering the kind of relationships people have with the institutions that make important decisions about their lives. The Trial captures the sense of helplessness, frustration, and vulnerability one experiences when a large bureaucratic organization has control over a vast dossier of details about one's life and on which decisions are made without the knowledge of the individual concerned."

In fact, it has been noted that the study of privacy [35] grew out of urbanization and the emergence of mass society. The fear was that citizens would pursue privacy at the expense of societal or communal participation. Privacy was negatively interpreted to reflect individual alienation and seclusion from public life. To this day, legitimate individual desires for solitude and privacy, whether online or offline, have been mistaken for secrecy and "hiding." Google Chairman Eric Schmidt, during an interview in 2009 with CNBC stated, "If you have something that you don't want anyone to know, maybe you shouldn't be doing it in the first place." This "nothing to hide" argument [36] erroneously suggests that privacy is something that only criminals desire, not law-abiding citizens. Nothing could be further from the truth. The society that Schmidt was referencing bears a close resemblance to a totalitarian state, not a free and democratic society.

In his book written in 1998, aptly entitled "The Transparent Society: Will technology force us to choose between privacy and freedom?" Brin [37] describes two 21st century cities. City number one utilizes ubiquitous technology as a surveillance tool to manage citizen behavior. In this Orwellian city, there is no privacy—it is trumped by a focus on security and public order. By contrast, city number two has the same technology and systems, but there are rules [38] in place that preserve private spaces and the citizens are the focal point for any benefits derived from the technology. "Guarantees of privacy, that is, rules as to who may and who may not observe or reveal information about whom, must be established in any stable social system. If these assurances do not prevail—if there is normlessness with respect to privacy—every withdrawal from visibility may be accompanied by a measure of espionage. For without rules to the contrary persons are naturally given to intrude upon invisibility."

Privacy and data protection will be the issues needed to be addressed in the evolution of a Cognitive City. As outlined earlier, privacy is an eternal value [39] in a free and democratic society. Already, public trust in the internet has been eroded to cause significant concern and several calls to action. The Global Commission on Internet Governance acknowledged that for the Internet to remain a global engine of social and economic progress, confidence must be restored. The Commission calls on the global community to build a new social compact with the goal of restoring trust and enhancing confidence in the Internet. One of the core elements in this new social compact [40] is that privacy and personal data protection be addressed by all governments and stakeholders as a fundamental human right.

3 Privacy by Design: An Essential Element for Building Cognitive Cities

This section outlines a 21st century privacy framework that is not limited by a zero-sum mindset but rather positions itself as a means to search for solutions that provide for both privacy and other societal objectives. When applied, citizens are presented with the best of both worlds, where difficult choices do not have to be made. If the cornerstone of the success of a Cognitive City are actively engaged citizens willing to participate as "human sensors" in all aspects of such a city's data ecosystem, we propose that a privacy framework will be essential for such participation. Privacy by Design (PbD) and the associated seven Foundational Principles presents such a framework.

3.1 Origins and Overview of Privacy by Design

The origin of Privacy by Design [41] grew out of the realization that the advances being made in the digitization of information and associated technologies as a result of the dawning of the Age of the Internet in the early 1990s, required a new framework. It was evident that such rapid advances in computing capability to process, copy and distribute data would contribute to a paradigm shift in the way we would need to think about preserving privacy—not just as a right, but as an eternal value. Privacy by Design advances the view that the future of privacy cannot be assured solely by compliance with regulatory frameworks; rather, privacy assurance must be embedded by design—ideally as an organization's or a technology's default mode of operation.

This paradigm shift includes the hallmarks of Privacy by Design which are: (i) taking a holistic, integrated approach to privacy and data protection measures by looking not only at policies, procedures and operational processes but also technology, networked infrastructures and physical design; (ii) expanding privacy beyond a regulatory framework of reactive enforcement after-the-fact, by pushing the need to proactively address privacy protective measures back to the conceptual or development stages; (iii) finding the means, where feasible, to make privacy assurance automatic or the default setting in any system or device; (iv) integrating and embedding privacy controls as essential design features; and (v) finding innovative solutions that foster the co-existence of privacy alongside other functional requirements.

Privacy is often viewed as an easy trade-off to attain other socially desirable, but competing goals. Zero-sum thinking is not only flawed—it is dangerous. It ignores the fact that technology can indeed be designed to provide a positive-sum, win/win outcome—the basis of Privacy by Design. Unlike some critics, who see technology as necessarily eroding privacy, we have long taken the view that technology is inherently neutral. We agree, however, that it is not the case when it is used to erode privacy—we must ensure that privacy becomes a central design goal in its own right. By this we mean building into technology the capability to achieve multiple functionalities—public safety and privacy, or using personal data within the constraints of privacy such that both businesses and users may benefit from its use. In this way one can literally transform technologies normally associated with surveillance into ones that are no longer privacy-invasive. This approach serves to minimize the unnecessary collection, use and disclosure of personal data, and promotes greater public confidence and trust in the specific technology and overall data governance.

What matters are the choices we make when designing and using these systems—ICTs can be privacy-invasive or privacy-enhancing, depending on their design. One of the first "privacy enhancing technologies" dates back as early as the 1970s with a focus on identity protection by anonymisation or pseudonymisation such as a 'mix networks' developed by Chaum [42] that provided anonymous and unobservable communications over a network. Such technologies [43–45] embody fundamental privacy principles by minimizing personal data use, maximizing data security, and empowering individuals to regain control over their personal information.

In the mid-90s, the idea of shaping technology according to privacy principles was discussed among Privacy and Data Protection Commissioners when the notion of embedding privacy into the design of technology was far less popular. At that time, taking a strong regulatory approach was the preferred course of action. Always a social norm, privacy has nonetheless evolved over the years, beyond being viewed solely as a legal compliance requirement, to being recognized as a market imperative and critical enabler of trust and freedoms in our present-day information society. With advanced digitization of data, networked infrastructure, social networking, the Internet of Everything and Big Data, it is now clear that the future of privacy cannot be assured solely by compliance with regulatory frameworks. Rather, privacy assurance must ideally become an organization's default mode of operation over three domains: (1) information technology; (2) accountable business practices; and (3) physical spaces and networked infrastructures.

In fact, this is exactly why the approach of designing privacy into the technology [30], by default, is necessary because "it is dangerously complacent to assume that social and organisational controls over accessibility of personal information are sufficient, or that intrusions into privacy will ultimately become acceptable when traded against potential benefits. Such a position could leave individual users with a heavy burden of responsibility to ensure that they do not, even inadvertently, intrude on others. It also leaves them with limited control over their own privacy."

The objectives of PbD—ensuring privacy and gaining personal control over one's information and, for organizations, gaining a sustainable competitive advantage—may be accomplished by practicing the following 7 Foundational Principles.

3.2 Privacy by Design Foundational Principles

The Privacy by Design framework consists of 7 principles [46] to be applied by organizations in a holistic, multi-pronged manner. They are not 'steps' to be taken, but rather a framework that expresses the full spirit of respect for privacy.

1. *Proactive not Reactive; Preventative not Remedial*: Whether applied to information technologies, organizational practices, or networked information ecosystems, PbD begins with an explicit recognition of the value and benefits of proactively adopting strong privacy practices, early and consistently, thereby preventing data breaches from happening in the first place. This implies: actions at the highest levels of organizational governance in order to demonstrate the value of privacy; an understanding of shared responsibility for privacy, characteristic of a culture of continuous improvement; controls in place to detect and mitigate privacy risks, well in advance.

2. *Privacy as the Default Setting*: This reflects the need to automatically protect personal information in any given IT system or business practice. To the degree possible, no action should be required on the part of the individual to protect their data—it should be built into the system, by default. Business and application owners can play a critical role in defining the necessary requirements for software engineers—such as, the importance of minimizing the collection and limiting the use of personal information to only that which is necessary for the purpose identified. Wherever possible, identifiability, observability and linkability of personal information should be minimized. In his book Code 2.0,[2] U.S. academic Lawrence Lessig wrote, "As the world is now, code writers are increasingly lawmakers. They determine what the defaults of the Internet will be; whether privacy will be protected; the degree to which anonymity will be allowed; the extent to which access will be guaranteed. They are the ones who set its nature."

3. *Privacy Embedded into Design*: Privacy must be embedded into technologies, operations, and information architectures in a holistic, integrative and creative way. Holistic, because additional, broader contexts must always be considered. Integrative, because all stakeholders and interests should be consulted. Creative, because embedding privacy sometimes means re-inventing existing choices because the alternatives are unacceptable. One of the tools developed to apply this principle is a Privacy Impact Assessment (PIA).[3]

4. *Full Functionality—Positive-Sum, not Zero-Sum*: Privacy and data protection are often positioned as opposites in a zero-sum manner; that is, as having to compete with or superseded by other legitimate interests, design objectives, and technical capabilities in a given domain. We reject that either/or, win/lose paradigm. When embedding privacy into a given technology, process, or system, it should be done so that all requirements are optimized. The notion that privacy requirements must be traded off against other interests (e.g. security vs. privacy or performance vs. privacy) is discarded as a dated formulation from the past. Innovative privacy solutions must prevail.

5. *End-to-End Security—Full Lifecycle Protection*: Strong security measures are essential to privacy, from start to finish—without strong security, end-to-end, there can be no privacy. This ensures that all data are securely retained, and then securely destroyed at the end of the data life-cycle, in a timely fashion. Applied security standards must assure the confidentiality, integrity and availability of personal data throughout its lifecycle including, inter alia, methods of secure destruction, strong encryption, and effective access control based on attributes [47], as well as logging methods.

---

[2]http://harvardmagazine.com/2000/01/code-is-law.html.

[3]Wright, D., Fin, R. and Rodriges, R. "A Comparative Analysis of Privacy Impact Assessment in Six Countries." (2013). Journal of Contemporary European Research, Vol 9, Issue 1, pp. 160–180.

6. *Visibility and Transparency—Keep it Open*: Visibility and transparency are hallmarks of a strong privacy program—one which will inspire trust in an organization's accountability and governance structure. This requires openness about an organization's policies and practices regarding personal information and its overall information management program allowing all stakeholders to provide the necessary oversight. Organizations are the custodians, not the owners, of our personal information. Audit trails are important to help users understand how their data is stored, protected and accessed.

7. *Respect for User Privacy—Keep it User-Centric*: This principle emphasizes the need for human-machine interfaces to be human-centered, user-centric and user-friendly so that informed privacy decisions may be reliably exercised. The concept of "user-centricity" has evolved into two sometimes contradictory meanings in networked or online environments. It contemplates a right of control by an individual over his or her personal information when online, usually with the help of technology. For most system designers, it describes a system built with individual users in mind, and which anticipates and addresses users' privacy interests, risks and needs. One view is libertarian (informational self-determination), the other is somewhat paternalistic. Privacy by Design embraces both understandings of user-centricity. Information technologies, processes and infrastructures must be designed not just for individual users, but also structured by them. Users are rarely, if ever, involved in every design decision or transaction involving their personal information, but they are nonetheless in an unprecedented position today to exercise a measure of meaningful control over those designs and transactions, as well as the disposition and use of their personal information by others.

Bellotti and Sellen [30] arrived at the same conclusion. In their study of "media spaces" such as EuroPARC's RAVE, a research laboratory designed to support distributed research collaboration, control (giving individuals the power to determine what information is shared and to whom) and feedback (notice to individuals on what information is being collected and to whom it is being disclosed) were found to be essential principles for designing in privacy into ubiquitous computing systems and influenced the design framework they developed for Computer Supported Cooperative Work technology (CSCW).

Similarly, business operations and physical architectures should also demonstrate the same degree of consideration for the individual, who should feature prominently at the center of operations involving collections of their personal data. The privacy interests of the end-user, customer or citizen are paramount. Where feasible, empowering data subjects to play an active role in the management of their own data may be the single most effective check against abuses of one's privacy and personal data.

4 Examples of Privacy by Design Approaches to Technologies Relevant to Cognitive Cities

For over a decade, the co-authors [46] have been involved in documenting not only what the concept of Privacy by Design actually means but also on the practical side, how the framework can be effectively applied. We acknowledge that much work has been done on privacy enhancing technologies [42–45] that would benefit the discussion on Cognitive Cities. For example, Oxygen is an MIT privacy project [48] where a range of context based information gathering technologies are under development. There, researchers are simultaneously co-designing in privacy regulatory features that address user control over their personal information.

In many cases, we focused on emerging technologies that invariably, when applied to a particular domain, had the potential to bring many new benefits and insights, but which equally raised privacy concerns. The technologies spanned a minimum of nine application areas such as: video surveillance cameras (CCTV); biometrics; smart meters; near field communications (NFC); RFID and sensors; geolocation advertising; wearables and wireless health care devices; and anonymous video analytics.

This section highlights three examples that showcase the value and applicability of taking a Privacy by Design approach to developing technologies and systems relevant to Big Data and the vision of a Cognitive City.

The first involves an ethical approach to analytics across large datasets (both structured and unstructured) to achieve real-time results. The second example deals with building privacy into a retail mobile location analytics platform that uses ubiquitous computing through sensor technology and WIFI/Bluetooth mobile device signals. The third example points to research on a concept of privacy-protective surveillance through the use of virtual cognitive agents and data analytics using homomorphic encryption.

4.1 Privacy by Design, Big Data and Context Computing

Context computing or 'sensemaking' systems are advanced approaches to Big Data, established to overcome the limitations of pattern-finding or data mining algorithms used by organizations. This new class of analytics [49] seeks to take new observations or transactions and integrate them with other pieces of data gathered through earlier observations or transactions, much in the same way one takes a jigsaw puzzle piece and locates its companions on the table. Context accumulation allows the process to occur fast enough to permit the user do something about whatever is happening while it is still happening. Unlike many existing analytic methods that require users to ask questions of systems, these new systems operate on a different principle: "the data finds the data, and the relevance finds the user."

Context computing may be viewed as an element of a cognitive system or a small subset of the cognitive suite of services and that helps to distinguish cognitive systems from current static systems. Hurwitz et al. [25] parsed out several elements of a cognitive system, including: (i) infrastructure and deployment modalities; (ii) data access, metadata and management services; (iii) the corpus, taxonomies and data catalogs; (iv) data analytic services; (v) continuous machine learning; (vi) hypothesis generation and evaluation; (vii) the learning process; and (viii) presentation and visualization services.

Cognitive systems are distinguishable from current computing applications because tabulating and calculating based on preconfigured rules and programs are replaced by the system's ability to [25]: (i) learn from experience with data/evidence and improve its own knowledge and performance without reprogramming; (ii) generate and/or evaluate conflicting hypotheses based on the current state of its knowledge; (iii) report on findings in a way that justifies conclusions based on confidence in the evidence; (iv) discover patterns in data, with or without explicit guidance from a user regarding the nature of the pattern; (v) emulate processes or structures found in natural learning systems (that is, memory management, knowledge organization processes, or modeling the neurosynaptic brain structures and processes); (vi) use NLP to extract meaning from textual data and use deep learning tools to extract features from images, video, voice, and sensors; and (vii) use a variety of predictive analytics algorithms and statistical techniques.

It is easy to see the impact on privacy of such a context computing system not to mention the security challenge. The fear is that the insights arising from such systems will be open to misuse by unauthorized individuals and that the system itself may be misused to further erode one's freedoms and liberty.

These seven features for next-generation sensemaking or context computing systems are a manifestation of Privacy by Design as articulated by the developer, Jeff Jonas [49] and his engineering team at the early stages of their work.

1. *Full Attribution*: Full attribution means recipients of insight from the sensemaking engine must be able to trace every contributing data point back to its source. When systems use merge/purge processing it becomes difficult to correct earlier mistakes (when a different assertion should have been made) as some original data has been discarded. Full attribution also enables system-to-system reconciliation audits of the data—particularly important when dealing with large information-sharing environments.

2. *Data Tethering*: Adds, changes and deletes occurring in systems of record must be accounted for, in real time, in sub-seconds. Data currency in information-sharing environments is important, especially where data is used to make important, difficult-to-reverse decisions that may affect people's freedoms or privileges. For example, if derogatory data is removed or corrected in a system of record, such corrections should appear immediately across the information-sharing ecosystem.

3. *Analytics on Anonymized Data*: An innovative technique enabling advanced data correlation while using only irreversible cryptographic hashes. This new

technique makes it possible for organizations to discover records of common interest (e.g., identities) across systems without the transfer of any personally identifiable information. This privacy-enhancing technology, known as "anonymous resolution" significantly reduces the risk of unintended disclosure while enabling technology to contribute to critical societal interests such as clinical health care research, aviation safety, homeland security and fraud detection. Reduction of risk without a material change in analytic results makes for a very compelling case to anonymize more data, not less.

4. *Tamper-Resistant Audit Logs*: Every user search must be logged in a tamper-resistant manner—even the database administrator should not be able to alter the evidence contained in this audit log. The question "Who will watch the watchmen?" remains as relevant today as when it was first posed in Latin two thousand years ago. People with access and privileges can, and do, occasionally look at records without a legitimate business purpose, e.g., an employee of a banking system looking up his neighbour's account. Tamper-resistant logs make it possible to audit user behavior.

5. *False Negative Favoring Methods*: The capability to more strongly favor false negatives is of critical importance in systems that could be used to affect someone's civil liberties. In many business scenarios, it is better to miss a few things (false negatives) than inadvertently make claims that are not true (false positives). False positives can feed into decisions that adversely affect people's lives—e.g., the police find themselves knocking down the wrong door or an innocent passenger is denied permission to board a plane. Sometimes a single data point can lead to multiple conclusions. Systems that are not false negative favoring may select the strongest conclusion and ignore the remaining conclusions.

6. *Self-correcting False Positives*: With every new data point presented, prior assertions are re-evaluated to ensure they are still correct, and if no longer correct, these earlier assertions can often be repaired—in real time. A false positive is an assertion (claim) that is made, but is not true; e.g., consider someone who cannot board a plane because he or she shares a similar name and date of birth as someone else on a watch list. Where false positives are corrected by periodic monthly reloading, wrong decisions can persist for up to a month, even though the system had sufficient data points on hand to know beforehand. In order to prevent this, earlier assertions need to be reversed in real time and at scale, as new data points present themselves.

7. *Information Transfer Accounting*: Every secondary transfer of data, whether for human consumption or to a tertiary system, must be recorded to allow stake-holders (e.g., data custodians or the consumers themselves) to understand how their data is flowing. In order to monitor information flows, information transfer accounting can be used to record both (a) who inspected each record and (b) where each record was transferred. This increases transparency into how systems are used. One day, it could enable a consumer, in some cases, to request an information recall. As an added benefit, when there is a series of information leaks (e.g., an insider threat), information transfer accounting makes discovery

of who accessed all records in the leaked series a trivial computational effort. This can narrow the scope of an investigation when looking for violating members within an organization.

4.2 Mobile Location Analytics

Analyzing consumer traffic patterns, whether in a shopping mall, stadium, airport, museum or any public space can provide important insights valuable to retailers, events management companies, flight authorities, to name but a few. Mobile location analytics (MLA) refers to a combination of technologies, including the use of sensors, that detect mobile device Bluetooth or Wi-Fi radio signals and then generate aggregate reports on the behavior of mobile device owners by using the device's MAC address as a unique identifier.

The privacy challenge for MLA and other sensor based applications, whether deployed in the retail, health or other private or public sectors, is, ironically, the very objective of ubiquitous computing.[4] The concept of ubiquitous computing is to embed technology in an unobtrusive manner into everyday objects to allow the seamless transmission and receipt of information from any other object or device. This very premise is one that permits the potential misuse of the technologies because of the lack of transparency and in turn, accountability to the individuals from whom the data is collected.

Information collected and processed by MLA technologies—frequency and times of visits, walking paths, dwell times—is highly revealing about an individual's lifestyle or habits. In the U.S., industry's response has been to introduce a Code of Conduct[5] (which aligns with PbD) for incorporating privacy considerations when implementing mobile location analytics solutions.

Already, these and other technologies such as infrared motion detection, thermal heat sensors, are being used in cities and neighborhoods around the world. Anonymous Video Analytics (AVA) [50] demonstrates how digital screen network operators are using pattern detection technology to understand viewing audiences while respecting consumer privacy. AVA is a technology typically used in the retail sector that scans real-time feeds from video cameras utilizing pattern detection algorithms to identify shoppers anonymously for the purpose of creating aggregate reports.

Efforts to design architectural elements and functional results consistent with a Privacy by Design approach to MLA [51] have been made. Some of the features consist of: MAC address pseudonymization; the use of randomization tables to reduce the risk of re-identification; ensuring the security of both data at rest and in transit through strong encryption protocols; aggregated reports not intended for

---

[4]https://www.ftc.gov/news-events/blogs/techftc/2014/02/my-phone-your-service.

[5]http://www.futureofprivacy.org/wp-content/uploads/10.22.13-FINAL-MLA-Code.pdf.

tracking or profiling individuals; providing clear notice to device owners about the data collection and uses; and persistent opt-out for consumers.

4.3   SmartData: A Cognitive Agent Approach to Data Protection

Earlier, we touched upon HDI as a natural next step in the way in which we relate to the evolution in technology. So while at one time, as users, we have been on the "outside looking in", perhaps we might extend this further than HDI to ask if the use of artificial intelligence and cognitive systems thinking may allow us to orchestrate data to protect itself while allowing for the openness and sharing required for unlocking the value of personal information. This involves the concept of intelligent or "smart agents" [52–54] that are evolved virtually into IT systems—creating SmartData (or personal avatars) that can think, understand, learn and remember the needs and privacy preferences of the individual to whom the data relates. The goal is to surpass current limited and brittle data protection methods by being able to respond to unforeseen situations, adapt to novel threats, and provide an accurate and nuanced representation of an individual's privacy and data security preferences.

This concept of a smart agent was extended to an application in the realm of intelligence-led surveillance. Privacy-protective surveillance (PPS) [55] uses modern cryptography, to ensure that (a) any personally identifying information (PII) on any unrelated individuals is not collected by the intelligence agency and (b) in transactions associated with targeted activity, PII and the metadata of additional "multi hop" connections will be encrypted upon collection, analyzed securely and effectively, and only divulged to the appropriate authorities with judicial authorization (a warrant). The objective is one where both privacy and public safety/security would be optimized.

Figure 1 below depicts the underlying architecture that incorporates three essential technological components. The first involves AI designed intelligent virtual agents whose purpose is to seek out, flag and then encrypt the personal information associated with pre-identified ("flagged") online suspicious activities or features. Secure multi-party computation methods (e.g. homomorphic encryption) are used to allow for 'private' interrogation and analysis of the online activity by way of probabilistic graphical models, to reduce false-positives and one's unnecessary disclosure of personal information. Not only does PPS limit its collection of data to "significant" transactions or events that are believed to be associated with terrorist-related activities, it analyzes that data wholly within the encrypted domain, thus providing additional assurance to customers that, in the off chance that their PII was collected by the system, no "prying eyes" would be able to record or monitor their actions from within the system. The long-term goal of PPS is to enable companies, as a public service to help counteract terrorism, to scan their own

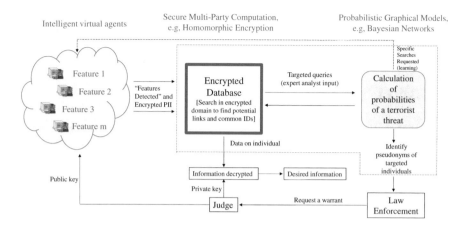

Fig. 1 Privacy-protective surveillance. *Source* Cavoukian, A. and El Emam, K. "Introducing privacy-protective surveillance: achieving privacy and effective counter-terrorism." IPC, Ontario. September 2013

databases, using PPS agents, and then turn over to law enforcement a copy of the encrypted files for anonymized analysis in a privacy protective manner.

5   Conclusion

In this chapter, the co-authors acknowledge the trajectory on which modern cities are moving and the enthusiasm that cognitive computing technologies bring to creating a symbiotic relationship between citizens, their ideal city and data. It is tempting to juxtapose privacy as an eternal human value or social norm alongside accrued societal benefits from anticipated free flow of personally identifiable information. The vision is clear—if we are to enlist the capabilities of cognitive computing and other innovative technologies in conjunction with a focus on the role of citizens as human sensors within such systems, then privacy and data security must become essential components in the design of such systems.

The willingness to share our personal information waxes and wanes and depends upon the context in which we find ourselves. It is this very willingness, based upon individual control, that lies at the heart of privacy and the choices we make about how much and to whom we wish to share the details of our lives and identities. In building Cognitive Cities, the timing is ideal to insert privacy into their development, given that we are at the beginning of the discussions. Moreover, using a Privacy by Design framework will encourage innovative design approaches to be taken—ones that will enable both privacy and the desired benefits of the systems integral to Cognitive Cities—a win/win proposition! Three examples demonstrate that it can be done. Organizations must seek ways to use the wealth of knowledge they have about citizens to provide better services to them in ways that increase trust not suspicion.

References

1. Mortier, R., Haddadi, H., Henderson, T., McAuley, D., Crowcroft, J.: Human-data interaction: the human face of the data-driven society (1 Oct 2014). Available at SSRN: http://ssrn.com/abstract=2508051 or http://dx.doi.org/10.2139/ssrn.2508051
2. Mostashari, A., Arnold, F., Mansouri, M., Finger, M.: Cognitive cities and intelligent urban governance. Netw. Ind. Q. **13**(3) (2011)
3. Sennett, R.: Open City. Retrieved from Richard Sennett (25 Oct 2013). https://www.richardsennett.com/site/senn/UploadedResources/The%20Open%20City.pdf
4. United Nations.: Department of Economic and Social Affairs, Population Division (2014). World Urbanization Prospects: The 2014 Revision, Highlights (ST/ESA/SER.A/352)
5. Coates, J.F.: Wild Ideas for Future Cities. Bridges—National Academy of Engineering Newsletter, **29**(4) (1999, Winter)
6. Washburn, D., Sindhu, U.: Helping CIOs Understand "Smart City" Initiatives. Forrester Research Inc, Cambridge Massachusetts (2010)
7. Kaltenrieder, P., Portmann, E., D'Onofrio, S., Matthias, F.: Applying fuzzy analytical hierarchy process in cognitive cities. ACM (2010)
8. Tusnovics, D.: Cognitive City: interdisciplinary approach reconsidering the process of (re) inventing urban habitat. In: Corp 2007 Conference, Vienna, (2007). http://www.cityfutures2009.com/program.html, Accessed from 11 June 2009
9. Finger, M.P.: From e-Government to e-Governance? toward a model of e-Governance. Electron. J. e-Government **1**(1), 52–62 (2003)
10. Oakley, K.: What is e-governance? Strasbourg: Council of Europe Integrated Project 1—Making democratic institutions work: e-governance workshop (2002)
11. Cavoukian, A.: Gov 2.0. Information and Privacy Commissioner's Office/Ontario, Toronto (2009)
12. Clarke, R.: Beyond the OECD Guidelines: Privacy Protection for the 21st Century (2000). Retrieved from Roger Clarke's website: http://www.rogerclarke.com/DV/PP21C.htm
13. NIST: Guide to Protecting the Confidentiality of Personally Identifiable Information (PII). National Institute of Standards and Technology, Washington, DC (2010)
14. Finn, R.L., Wright, D., Friedewald, M.: Seven types of privacy. In: Gutwirth, S., Poullet, Y. (eds.), European Data Protection: Coming of Age (5th edition of Computers Privacy and Data Protection (CPDP) ed.). Springer, Brussels (2013)
15. Nissenbaum, H.: A contextual approach to privacy online. Daedalus **140**(4), 32–48 (2011)
16. IPU Assembly Standing Committee on Democracy and Human Rights: Democracy in the Digital Era and the Threat to Privacy and Individual Freedoms. 132nd IPU Assembly—Interactive Debate. Interparliamentary Union Assembly, Hanoi (2015)
17. Cavoukian, A.: TTC Investigation Report. Information and Privacy Commissioner's Office, Toronto, Ontario, Canada (2006)
18. Citron, D., Gray, D.: The right to quantitative privacy. Minn Law Rev **98** (2013, March 5)
19. Cavoukian, A.: Surveillance, Then and Now: Securing Privacy in Public Spaces. Information and Privacy Commissioner's Office of Ontario, Ontario (201)
20. Gutwirth, S.: Privacy and the Information Age. Rowman & Littlefield, Lanam (2002)
21. Cavoukian, A.: A Primer on Metadata: Separating Fact from Fiction. Information and Privacy Commissioner's Office, Toronto, Ontario (2013)
22. Kuneva, M. Keynote Speech. Roundtable on Online Data Collection, Targeting and Profiling. Brussels, 31 March 2009

23. World Economic Forum: Personal Data: The Emergence of a New Asset Class. Switzerland, Geneva (2011)
24. Feldman, S., Reynolds, H.: Cognitive Computing: A Definition and Some Thoughts. KM World, November/December (2014) [Vol. 23, Issue 10]
25. Hurwitz, J., Kaufman, M., Bowles, A.: Cognitive Computing and Big Data Analytics. Wiley, Somerset (2015). http://ryerson.eblib.com/patron/FullRecord.aspx?p=1895129. Accessed 19 Oct 2015
26. Dey, A.K., Salber, D. Abowd, G.D.: A conceptual framework and a toolkit for supporting the rapid prototyping of context-aware applications, anchor article of a special issue on context-aware computing. Hum. Comput. Interact. (HCI) J. 16(2–4), 97–166 (2001)
27. Fischer-Hubner, S., Hoofnagle, C., Krontiris, I., Rannenberg, K., Waidner, M. (ed.) Online Privacy: Towards Informational Self-Determination on the Internet. Manifesto from Dagstuhl Perspectives Workshop, 6–11 Feb 2011
28. Dey, A.K., Hakkila, J.: Context-awareness and mobile computing. In Lumdsen, J. (ed.) Handbook of Research on User Interface Design and Evaluation for Mobile Technology. Idea Group, Inc. Chapter 13 (2008)
29. International Working Group on Data Protection in Telecommunications (IWGDP-T): Working Paper on Big Data and Privacy: Privacy principles under pressure in the age of Big Data analytics (May 2014)
30. Bellotti, V., Sellen, A.: Design for privacy in ubiquitous computing environments. In: Proceedings of the Third European Conference on Computer-Supported Cooperative Work. 13–17 September 1993 Milan, Italy, pp. 77–92. Springer Netherlands (1993)
31. Pekarek, M., Roosendaal, A., Sluijs, J.: Surveillance as a Service? On the use of surveillance data for administrative purposes. In: Gutwirth, S., Leenes, R., DeHert, P., Poullet, Y. (eds.) European Data Protection: Coming of Age, pp. 347–365 (2013)
32. Zittrain, J.: The Future of the Internet and How to Stop It. Yale University Press, New Haven (2008)
33. Solove, D.J.: The Digital Person: Technology and Privacy in the Information Age. NYU Press (2004). Available at SSRN: http://ssrn.com/abstract=609721
34. Whitehead, J.W.: Whitehead: Kafka's America: Secret Courts, Secret Laws and Total Surveillance. The Rutherford Institute (2013)
35. Zureik, E., Stalker, L. L.H., Lyon, D., Smith, E., Chan, Y.E.: Preface. In: Zureik, E., Stalker, L.L.H., Smith, E., Lyon, D., Chan, Y.E. (eds.) Surveillance, Privacy, and the Globalization of Personal Information: International Comparisons (p. ix–xii). McGill-Queen's University Press (2010). http://www.jstor.org/stable/j.ctt1q606m.3
36. Abdo, A.: You may have 'nothing to hide' but you still have something to fear. (ACLU, Ed.) California, USA (2013). https://www.aclu.org/blog/you-may-have-nothing-hide-you-still-have-something-fear. Accessed from 2 Aug 2013
37. Brin, D.: The Transparent Society: Will Technology Force us to Choose Between Privacy and Freedom?. Addison-Wesley, Reading, Mass (1998)
38. Schwartz, B:. The sociology of privacy. Am. J. Sociol. 73 (1968)
39. Schneier, B.: The Eternal Value of Privacy. Retrieved from Wired Commentary (18 May 2006). http://www.wired.com/print/politics/security/commentary/
40. Global Commission on Internet Governance: Toward a Social Compact for Digital Privacy and Security. Statement by the Global Commission on Internet Governance. 15 April 2015. Sourced 17 Oct 2015. https://www.ourinternet.org/publication/toward-a-social-compact-for-digital-privacy-and-security/
41. Cavoukian, A.: privacy by design—Part III. In: Gutwirth, S., Leenes, R., Hert, P.., Poullet, Y. (ed.) European Data Protection: Coming of Age. Computers, Privacy and Data Protection (Conference). (2013)
42. Chaum, D.L.: Untraceable electronic mail, return addresses, and digital pseudonyms. Commun. ACM, 24(2), 84–90 (1981). ISSN 0001-0782. doi:http://doi.acm.org/10.1145/358549.358563

43. Danezis, G., Domingo-Ferrer, J., Hansen, M., Hoepman, J.-H., Tirtea, R., Schiffner, S.: Privacy and Data Protection by Design: From Policy to Engineering. ENISA, Heraklion, Greece (2014)
44. Enterprise Privacy Group (EPG): Privacy by Design: An overview of Privacy Enhancing Technologies. Odiham, UK (2008)
45. Shen, Y., Pearson, S.: Privacy Enhancing Technologies: A Review. HP Laboratories, Bristol, UK (August 2011)
46. Cavoukian, A.: Operationalizing Privacy by Design: A Guide to Implementing Strong Privacy Practices. Information and Privacy Commissioner's Office, Toronto, Ontario, Canada (2012)
47. Cavoukian, A., Chibba, M., Williamson, G., Ferguson, A.: The Importance of ABAC (Attribute Based Access Control) to Big Data: Privacy and Context. Privacy and Big Data Institute, Ryerson University, Toronto, Canada (2015)
48. Ackerman, M., Darrell, T., Weitzner, D.J.: Privacy in context. Hum. Comput. Interact. **16**, 167–176 (2001)
49. Jonas, J., Cavoukian, A.: Privacy by Design in the Age of Big Data. Information and Privacy Commissioner, Toronto, Canada (2012)
50. Intel Corporation, and Information and Privacy Commissioner/Ontario: Anonymous Video Analytics (AVA) Technology and Privacy: How Digital Screen Network Operators are Using Pattern Detection Technology to Understand Viewing Audiences while Respecting Consumer Privacy. Information and Privacy Commissioner of Ontario, Toronto, Ontario, Canada (2011). http://www.deslibris.ca/ID/229874
51. Cavoukian, A., Bansal, N., Koudas, N.: Building Privacy into Mobile Location Analytics (MLA) Through Privacy by Design (2014). http://alltitles.ebrary.com/Doc?id=10874151
52. Harvey, I., Cavoukian, A., Tomoko, G., Borrett, D., Kwan, H., Hatzinakos, D.: SmartData: Privacy Meets Evolutionary Robotics (2013). http://site.ebrary.com/id/10682294
53. Tomko, G.J., Kwan, H., Borrett, D.: SmartData: The Need, the Goal, the Challenge. University of Toronto, Identity, Privacy and Security Institute (2012)
54. Tomko, George, Borrett, Donald, Kwan, Hon, Steffan, Greg: SmartData: make the data "think" for itself. Identity Inf. Soc. **3**(2), 343–362 (2010)
55. Cavoukian, A., El Emam, K.: Introducing Privacy-Protective Surveillance: Achieving Privacy and Effective Counter-Terrorism. Information and Privacy Commissioner, Toronto, Canada (2013)

Towards the Improvement of Citizen Communication Through Computational Intelligence

Valerio Perticone and Marco Elio Tabacchi

Abstract When dealing with problems that arise from collective sharing of resources in metropolitan areas (i.e., energy, pollution, traffic, health) most of the interaction between citizens and local governance is usually carried out through the use of natural languages. Digital technologies allows smart cities residents to communicate with a broad range of experts (e.g. bureaucrats, legislators, urbanists, etc.) that routinely use technical terminology seldom accessible to the layperson, or linguistic styles that are not immediately understandable. Although information technology should encourage citizen participation in governance at many levels, the different levels of knowledge possessed by the actors can lead to incomprehension, as well as social exclusion. Computational Intelligence approaches can be used in order to alleviate such difficulties and improve the efficiency of communication through automation and collective cognitive systems. In this paper we discuss how use of techniques such as CWW and Fuzzy Classifiers can be beneficial toward the reduction of the communication gap between citizens and government.

1 Introduction

Natural language is an integral part of the interaction between citizens and policy makers, including various forms of local and regional government, the public and private health system, service providers and all the entities that contribute to the definition of a correct and fruitful relationship between the people intended as individuals and their collective instantiation. This is as true, if not more, today as it was in the past, as the number and quality of resources that are shared among

V. Perticone · M.E. Tabacchi (✉)
DMI, Università Degli Studi Di Palermo, Palermo, Italy
e-mail: marcoelio.tabacchi@unipa.it

V. Perticone
e-mail: valerio.perticone@unipa.it

M.E. Tabacchi
Istituto Nazionale Di Ricerche Demopolis, Demopolis, Italy

© Springer International Publishing Switzerland 2016
E. Portmann and M. Finger (eds.), *Towards Cognitive Cities*,
Studies in Systems, Decision and Control 63, DOI 10.1007/978-3-319-33798-2_5

citizens, especially in metropolitan areas, is constantly increasing and expanding, carrying with it a number of new problems and aggravating old ones. In a connected society, the problems brought out by the inevitable increase in complexity can be mitigated by the diffused use of information technology, but sometimes the introduction of easier-to-use and easily available communication means brings around as an unintended consequence a marked worsening of the usefulness of communication, as well as disaffection by the users—especially when dealing with special categories of citizens, such as the elderly, the disabled or members of the younger generation. The ubiquity of communication means (mobile phones, tablets, powerful net- and notebooks), the availability of fast and resilient networks in all but the most rural areas, the diffusion of communication points as a commodity service in a large part of the actual locations managed by the retail industry, a legislative and political push toward a complete digitalisation of the administrative process, the switch to digital in order to push for cost savings by classic and advanced service providers (utilities, media, parcel delivery), a number of development toward the concept of smart city, all these factors combined have created the perfect background to usher citizens in a new era of instant, direct, traceable communication with institutions and commercial parties alike. This dream scenario is fast approaching, as the numerous technical hurdles are solved or bypassed at an accelerating pace, but still a significative barrier exists: language. The problem of language in citizen communication is twofold: a strictly technical aspect pertains the automatic or semi-automatic translation of information generated by systems or collation of systems and based on collection or manipulation of raw data into some aggregated form that can be perused by the general public. In this context the attention of researchers has been devoted to problems such as general aspects in data analysis, big data generation, retrieval and association, contextual text analysis. While many computational intelligence methods have been successfully applied to a satisfactory resolution of such conundrums, the present article will not deal with is aspect, beside some passing remark in the technical section.

We are more interested in discussing the more strictly cognitive aspect of the dilemma, the one that arises when the technical language has to be translated into bits of information that are easily digested by a population that has more and more lost its ability for a critical examination of long and complex texts. The present technological infrastructure (as the proximate future will), once again with a prevalence in cities and metropolitan conurbations, allows a vast number of people to open communication channels with different policy makers and administrators; citizens discuss policies and quality of service with a broad range of experts, such as bureaucrats, legislators, urbanists and architects, physicians, managers, and the experts routinely use a lexicon and technical terminology seldom accessible to the layperson, when not deliberately obfuscate and incomprehensible, and linguistic styles that are not immediately understandable or decodable without access to further corpuses or external consultation. Information technology has among their priority the encouragement to participation in governance at many levels and with different aims, but due to the different levels of knowledge possessed by the actors, a problem that is rooted not in technology itself but in the typical approach to system development and deployment, such distance becomes incomprehension, and more often than not even digital social exclusion—a phenomenon that appears to cross the boundaries of age, education and wealth and seem more connected to the intimate relationship of the individual with the use of technology and the allocation of time resources to different activities, encompassing a large number of actors and situations. The most significative themes of social life and resource sharing in a community, such as energy production, distribution and consumption, pollution reduction and control, public and private transport, healthcare and general well-being, are directly impacted by the ability of the layperson to understand what is said and discussed in the public debate, and to express informed opinion based on reasoning and feelings in a timely and punctual manner, and developing and deploying communication systems that keep this concept at their foundation is of vital importance to increase democratic participation and fight the current trend of disaffection towards politics, both at the national and local level.

Enters computational intelligence, and more specifically the methodologies that stem from developments in fuzzy logic, fuzziness and uncertainty and incomplete information treatments: Computing with Words, Fuzzy Classifiers, Fuzzy Text Analysis, Metaheuristics. The premise of such techniques is clear-cut: the requisite for extreme precision in measurement that has characterised most of the engineering approach is relaxed in favour of a description that is more akin to the one of natural sciences, and that comes more from observation than modelization; in the words of Lotfi A. Zadeh, "a coalition of methodologies which are drawn together by a quest for accommodation with the pervasive imprecision of the real world" [47]. As we wrote elsewhere [37, 37], "fuzzy sets deal with ordinary language; they are mathematical entities contextually reflecting collectives, and that are mathematically modelled by their membership functions. They meant to pass from an old world of exact thinking represented by sets, to a new world of inexact thinking represented by them." It is easy to see that language has followed, in a much greater span of time and form a wider audience, the same path; the power of expression has always taken precedence over the formalisms of the grammar, which can be considered an ex-post solution to a homogenization problem more than a set of precise, generative rules. The use of technical language is rooted in dictionaries, but its expressivity also comes from a delicate bending of the rules in order to be more akin to the external reality it is willing to represent. Moreover the continually changing external situation is reflected in its liquidity, and the extreme complexity of the reality that has to be described insures that part of the necessary information in the communication game is always hidden or difficult to reach.

We maintain that this isomorphism has to be exploited by employing techniques from computational intelligence to improve citizen communication and elicit from individuals a better expression of sentiments, feelings and opinions about the political and social actions and consequences of policymaking, the quality of service from utilities and providers, and more in general the communication act in the reference communities. This is clearly a grand scheme, but our opinion is that in order to achieve the goal of a really smart city, when communities are more than just the sum of interests expressed by their components, this will be an unavoidable path, and the earlier we start adopting such methodologies, the better. Work has already been done in this direction, and we will comment in more detail in the next section. Scope of this paper is to give a hint of the methodologies that seem more suited and beneficial toward the reduction of the communication gap among citizens and between them and policymakers and experts, discussing both possible future applications and instance of successful previous applications to cognitive similar problems, and to present two possible examples of such applications to real world problems.

The rest of this paper is structured as follows: the next section will succinctly describe a number of specific techniques from the arsenal of computational intelligence. Such techniques have been chosen, among many others, as in our opinion they are the one that will be needed for a fruitful development of citizen communication systems based on natural language in the near future. Along with the brief description of their working mechanisms—we will give references for a more detailed description of each one—we will briefly highlight how such techniques can be fruitfully used toward our scopes, and a number of instances where they have been successfully applied in the solution of problems related to citizen communication or smart city development and deployment. The two following sections will deal with specific examples of citizen communication problems, and describe possible solutions based on computational intelligence methods. In particular we will deal with the problem of patient-physician communication in e-health, and describe a system to help patient in decoding interactively information contained in a dossier. In the second example we will discuss the problem of sentiment and consent expression as it is applied to collective and shared decisions, and a method for successful decoding of such information based on fuzzy text analysis. Conclusions and an outline for further work will follow.

2 Computational Intelligence Methodologies for Citizen Communication

Computational Intelligence methodologies are nowadays constantly employed in the solution and optimisation of problems that are cognitive bound (see e.g. [11, 13, 23]). Many such methodologies, especially the ones from the tradition of soft computing (metaheuristics, fuzzy systems), are in fact a natural match for cognition, as they are based on the fundamental idea that human cognition as a process is not based on symbol manipulation using strict rules, but has evolved to take into account a reality whose sheer complexity rapidly becomes intractable if not by acknowledging a constant selection of information, the uncertain nature of communication, and subjective probability as a rule of behaviour. In this section we briefly discuss some of the computational intelligence methodologies that can be usefully implemented in order to build systems that support citizen communication in a native and intuitive fashion, along with some example of implementations of such methodologies toward the solution of problems inherent to the development and deployment of smart cities and smart communities.

2.1 Metaheuristics

Metaheuristics is a general term indicating a class of algorithms that are designed to provide solutions to problems that are solvable in theory but not in practice in the sense that we know the steps needed to solve them, but the time it will take to solve them is not on the human scale. Problems of the NP class, for which known algorithmical solutions exists, but the time that is necessary to complete the execution of such algorithms becomes so high that no computer based on the present architectures can effectively run them as soon as the size of the problem rises, as well as optimisation problems involving uncertainty, incomplete or imperfect information, are examples of the kind of situations in which the application of metaheuristics can be beneficial. Metaheuristics essentially have in place two kinds of tradeoffs: the first is concerned with the exactness of solution. Since by definition an exact solution is not possible, metaheuristics concentrate on finding a suboptimal, approximate solution, which is not perfect but "good enough" for the problem at hand. The second has to do with the methods implemented to find the suboptimal solutions: metaheuristics do away with careful planning and systematisation of the parameters by making few assumptions about the optimization problem being solved and relying instead on a wide exploration of the space problem, often helped by spanning algorithms that take advantage of sheer computing power and the availability of big data, plus intelligently sampling a subset of all the possible solutions helped by casual choices. These two aspects make metaheuristics quite in place with cognitive endeavours, as they both recall modalities that are proper of the human reasoning: the idea of settling for a solution that may not be the best possible but that satisfies other, more stringent requirements (attainability, speed, practicality), and modalities that considers exploration, multiple tries, trial and error, collaboration and cooperation, further refinements.

Popular metaheuristics that are used in the context of smart cities development and communication are Simulated Annealing, Tabu Search, Local Search, Variable Neighborhood Search, Ant colony/Particle Swarms, Evolutionary Computation, and Genetic Algorithms, and such methodologies, which along fuzziness consti-tutes the bulk of soft computing, have recently seen a renewed interest, as witnessed by their popularity also in non specialistic press. This for at least two reasons: the fact that there is now sufficient distributed computing power to let everyone play with the concept (as metaheuristics are born to accommodate parsimony, something that explains the immediate popularity of methods such as Particle Swarms or GAs, but real, useful applications are in need of the kind of computing power only now available at a large scale); and the acceptance of these methodologies in ICT and education. Many lecture courses in engineering and computer science curricula deal with metaheuristics, and popular scientific libraries implement most of them out of the box, allowing non-experts and experts alike to implement and improve on what exists. Furthermore the existence of many graphical tools (especially in simulation) have attracted to metaheuristics less technically inclined researchers from humanities, strengthening their role as a bridge between different souls of cognition.

Metaheuristics can solve optimization problems related to urban transportation [30]: for example choosing an appropriate route from two points on a map (a departure point and a destination) for a number of vehicles as the road traffic conditions by choosing an appropriate route for a number of vehicles as road traffic conditions may rapidly change (e.g. after a road incident). The performance of the different metaheuristic in a real road networks also lays on the availability of information about traffic conditions. The huge amount of traffic data, gathered through sensors, can be analyzed using Soft Computing techniques applied on geographic information (Geospatial Analysis) [10]. A review of methods resulting from the field of agent and multiagent systems shows that these rapidly emerging techniques have been applied to many aspects of traffic and transportation systems because deals with the uncertainty in dynamic changing environments [5]. These methods can be used also for planning the adoption of strategies to promote sustainability in smart cities, analyzing prices, benefits and tradeoffs [22].

2.2 Computing with Words

Among the fuzzy methodologies mostly concerned with the cognitive effort made by humans in order to achieve linguistic or verbal communication, Computing With Words (CWW) is one of the most innovative ideas of the last decades, a visionary (in the good sense of the word) concept able to connect very distant concepts in a common framework. In assessing how such methodology is intrinsically connected to cognition, Lotfi A. Zadeh, the father of Fuzzy Logic, wrote in his introduction to the CWW paradigm that "The role model for CWW is the human mind" [46].

CWW provides a fusion between natural language, and specifically its verbal characteristics, and computation using fuzzy variables. In CWW, the basic information to be manipulated consist of a collection of propositions expressed in a natural language. While standard computation (of the von Neumann type) usually is based on input data, some kind of functions (usually realised as programs) that manipulate such data, and the output from such functions, in CWW our basic units, which are fuzzy constraints of a variable, called granules, and describe premises, some fuzzy constraint propagation, and conclusions. From the premises answers to a query expressed in a natural language are to be inferred, and the logical part of the problem is to derive such conclusions starting by the premises and using propagation. The presupposition of CWW is to replace (back again) numbers by words, in a reversal of what has been a longstanding principle of science, the arithmetization of everything: again with Zadeh, "in coming years, computing with words is likely to emerge as a major field in its own right. In a reversal of long-standing attitudes, the use of words in place of numbers is destined to gain respectability. This is certain to happen because it is becoming abundantly clear that in dealing with real-world problems there is much to be gained by exploiting the tolerance for imprecision. In the final analysis, it is the exploitation of the tolerance for imprecision that is the prime motivation for CWW."

The act of communicating verbally is an inherent cognitive activity, while computation seem to be a more typical application of the automatic enterprise. Turing was among the first to acknowledge the relationship between such approaches, with a work of systematization and simplification that reduced number to symbols, and their combination in order to attain a significant to basic operations [40]. Such work was extended by Chomsky and his contemporary, who tried to incorporate the idea of language in the same framework, and to derive rules for generation and manipulation [6–8]. CWW carries this idea a bit further, by recognising that the failure of theories of language based on the symbolic manipulation in reproducing human performance was mainly due to inherent characteristics of the language itself, linked to imprecision, incomplete information and the wiggle room a natural language has to create and maintain in order to be able to express an infinite set of nuances and intended meanings [21].

Among the applications of fuzziness to cognitive problems CWW has remained a bit behind the methodologies that will be illustrated in the next subsections, perhaps due to the lacking of a coherent formalisation. But where such applications were effectively developed, their elegance and simplicity demonstrated the powerfulness of the instrument and the high degree of proximity to cognition that are ineludible if we want to build citizen communication support in our systems.

CWW can be used for describing air quality in linguistic term and classifying pollution [45]. Measurement and control of polluted air can reduce respiratory diseases in urban areas and improve the quality of life. Another possible application of CWW is the analysis of linguistic phrases describing symptoms for diagnostic assessments of medical conditions [3].

A review of the applications of CWW to Decision Support Systems can be found in Martínez et al. [29]. As noted by the authors, Decision Making is an inherently human activity, usually modelled using linguistic information by the experts involved in the decision settings when qualifying performances related to human perception, using words from natural languages instead of numerical values. This echoes the problems incurred in citizen communication when experts are involved, and the solutions reviewed in the paper can be adapted to such problem with the same good performance obtained by linguistic computation.

We are convinced that further steps in the direction of a solid implementation of CWW would carry advantages in other endeavours relative to the development of smart cities in general, and that such approaches can be beneficial to sentiment and feeling expression and community sharing of information.

2.3 Fuzzy Classifiers

Classifiers are a staple of Computational Intelligence, as they allow making sense of the most diverse information, and are at the base of any systematisation of information. What fuzzy classifiers add to the mix is their ability to deal with the classification problem by stressing the significance of linguistic variables. While conventional classifier are mostly sterile from the cognitive point of view—they do their job well but offer no further insight on the process of classification that can be usefully employed at the cognitive level, fuzzy classifiers are widely employed because of their capability to built a linguistic model interpretable to the users and as they give the possibility of fusing information from different sources, as the one coming from expert knowledge and information coming from mathematical models or empirical measures. Another significative advantages of these systems is the high interpretability of the output model, which consents to add a semantic cognitive layer to its results. Fuzzy Rule-Based classifiers, which are a further specification of general Fuzzy Classifiers, have been successfully applied to pattern classification problems, as they provide a good platform to deal with uncertain, imprecise or incomplete information often present in cognitive system, and are an effort to reconcile the empirical precision of traditional engineering techniques and the cognitive interpretability of inputs and outputs that is a staple of Computational Intelligence.

The applications of Fuzzy Classifiers to smart cities development are too numerous to list. Examples include economical planning [2]: seeing as Fuzzy Rule-Based System can handle the complexity of financial management, one of the most important management tasks of city government, and resource management [1], predicting consumption values and making predictions about usage fluctuations of a specific asset (i.e., water).

2.4 Fuzzy Ontologies

In order to build a system that can support reasoning and the use of natural language, a structure that allows to save and retrieve information about the state of the word, plus a facilitator for making connections between such information in a natural way, is needed. A successful example of such structure in the domain of knowledge representation and reasoning are ontologies, but traditional ontologies are limited by the fact that the linking between entities is usually defined as the existence of a specific property in binary mode: the property either is there—and it links objects in a complete way—or is not. This creates representations in which subtle nuances, graded properties and the general uncertainty that characterises the human experience is invariably simplified, at the point of being not a useful representation anymore. Fuzzy-based ontologies are well versed to represent the implicit informations in a complex ontologic system, as the relationships between objects, entities are now represented by fuzzy quantities, often expressed using linguistic quantifiers. This generates an ontology that not only is a better representation of reality, but is also more clear to read and understand, and can be easily manipulated to incorporate updates in a timely manner, another potential drawback of classical ontologies, where it is difficult to pinpoint the exact moment where a relationship is formed or rescinded in a complex system. Fuzzy ontologies are not only ideal for knowledge storing and retrieval, but can also play an important part in Theory of Concepts, as they supply a compatibility layer in the framework of an intensional approach, another limit of traditional ontologies. All these characteristics make the use of fuzzy ontologies a boon for citizen communication, as in order to get to the semantic of natural language, a number of assumptions about the non-expressed part of the message have to be made and stored in an ontology.

Applications in which fuzzy ontologies are employed in the domain of collective communication and development include weather forecast, an environment where there are inevitable uncertainties and inconsistency associated with knowledge of natural phenomena [39], contexts that use Similarity Reasoning, for example in crisis and emergency management scenarios [18], and other social interactions [36]. In this context the idea of Fuzzy cognitive maps, which is another fuzzy methodology with roots in cognition, should be mentioned for its contributions to the handling of knowledge (see e.g. [24]).

In medical domain the construction of fuzzy ontologies can support health professionals for identify useful documents in vast source of knowledge because the increasing volume of available information make difficult find appropriate data in different corpora pertaining to patients' healthcare needs at the time of clinical point of care [33].

3 Possible Applications in Citizen Communications: E-health and Citizen Empowerment

One aptly fit example of citizen communication that can benefit from automated systems based on computational intelligence methods is the exchange of medical texts for patient's information and reference. Medical texts are infact usually written by professionals from the field, such as physicians, medical and academic researchers, who are keen to use their own language and communication style, both for the purpose of intra-communication clarity and also for a number of reasons that are not directly related with text comprehension by the final users of such texts— e.g. insurance or legal reasons. Once produced, such texts are read by exactly those final users which are not always in the main concern of the professional, and have no specialist skills and vocabulary as them, which often brings to problems in communication and reciprocal trust [27, 44]. This is especially true for medical reports or diagnoses, where a correct understanding by the users is essential in order to guarantee informed consent, and the choice of a therapy that could better suit the needs and the desires of the patients.

Efforts to reduce language complexity in order to render citizen communication easier to understand for a vast portion of citizens are already in force elsewhere: see e.g. citizen communication simplification project by the US government at Plain Language.gov [35] which uses "a kind of language that audience can understand the first time they read it", or Simple English Wikipedia, another example of content citizens' peruse, developed using only a restrict lexicon of the most common English words [43].

In this vein it should be also possible to reduce the complexity and variability of medical terms and text using a similar effort, but this would require the medical class to coordinate a specific effort starting at the academic level, and not everyone would accept the idea of a simplification of the language at source in fear of oversimplification, loss of important time in producing diagnoses or a significative reduction of expression power which is guaranteed by the medical jargon. For these reasons an ideal compromise would be an automated system that supplemented the informed reading of medical texts written in traditional form (of which a large corpus actually already exists) by translating all technical terms in their variability to terms that are more uniform and understandable, and adding further information in plain text so to support the uninformed reader in comprehending the exact meaning of the text and formulating eventual questions and request for clarifications. This way the user will not be requested to acquire a vast knowledge before being able to understand a diagnosis, but the knowledge itself would be brought to the user in an automated and punctual manner, in order to allow decoding of such texts. A further benefit from such system would be in its ability to become an instructional item, able to educate the professional to a more standardised and clear use of the technical language, by helping self reflection and checking for consistence. In this context CWW and fuzzy ontologies, a seen in the previous section, can be fruitfully used to analyze and enrich medical text, and facilitate citizen communication with the medical profession.

3.1 Medical Vocabularies and Dictionaries

We started with an investigation of the classical tools used from experts and laypersons to understand medical text. Medical vocabularies are selective lists of words and phrases used in the medical field. Usually they are created for the professionals and, for this reason, they contain all the technical terms used in the medical field. Citizens with no specific technical background rarely use such resources, confining their utilization to the situations in which specific terms cannot be found in consumer dictionaries or when more technical information is needed.

The 'Unified Medical Language System (UMLS)' is a large collection of multilingual vocabularies that contains information about biomedical and health related concepts created and maintained by the US National Library of Medicine [41]. It uses the idea of a 'Concept Unique Identifier—CUI' which is an unique identifier that links specific terms to a single concept) to create a mapping among these vocabularies and thus allows translation among the various terminology systems. The integration of these systems can be viewed also as a comprehensive thesaurus and ontology of health and biomedical concepts.

As for the counterpart to this problem, it is well known that consumer terms are not well covered by the existing medical dictionaries, as for a choice they tend to favor the lexicon from health professionals [25]. Indeed, expressions used by consumers to describe health-related concepts and relationships among such concepts frequently differ on multiple levels (i.e., syntactic, conceptual and explanatory) from those of professionals. To counteract this tendency, a number of specific consumer health vocabularies (CHVs) have been developed in the previous years, with the aim of translating terms and adapting concepts from the medical lexicon to their equivalent in the common parlance. Such vocabularies can be effective in translating technical terms from the domain of medicine to words that a citizen with no previous specific knowledge can relate to simple concepts and situations [48].

One outstanding example of such vocabularies is the 'Open Access Collaboratory Consumer Health Vocabulary (OAC-CHV)' created and maintained by the Consumer Health Vocabulary Initiative [9]. OAC-CHV structure is formed by a relationship file, that creates links between terms that are commonly employed for description of medical facs intended for the mass consumption to the specific counterparts that form the terminology in UMLS. Focus of OAC-CHV are expressions and concepts that are somehow considered when communication between patient and practitioner occurs.

3.2 *Fuzzy Automatic Comprehension of Medical Texts*

Using informations from vocabularies and dictionaries we can build a fuzzy automatic system for increase the comprehension of texts. The system takes as input an arbitrary text and, using the vocabulary, extracts all the technical terms in the area of the chosen subject and connects each term to its equivalent consumer terms using the ontology built using the thesaurus. The research of related terms is not restricted to technical words that have a consumer translation but also extended to synonyms of terms (technical or consumer, if applicable) and use information from the description retrieved by a consumer dictionary. This definition will also be processed by the whole system and transformed in an annotated hypertext that highlights the technical terms so to allow the user a deeper analysis and navigation.

To improve the comprehension of health procedures, our system uses an internal dictionary based UMLS. As a thesaurus, OAC-CHV is used. The mapping from UMLS to OAC-CHV is accomplished by means of the Concept Unique Identifier (CUI) when available [26]. Mapping is taken care of considering the fuzzy degree of significance for each term, calculated by building a fuzzy ontology (built by weighting the contribution of the term in each snippet of related text, and recreating OAC-CHV links in a fuzzy fashion using FuzzyOWL2).

In order to verify the effectiveness of our system and evaluate its usefulness in the citizen communication domain when dealing with health and medical issues we have employed a number of documents of different kind, such as medical statements and diagnoses and online medical reports. The system has proved particularly effective in the simplification of results from diagnostic procedures such as MRI and EEG, where the disambiguation and clarification of technical terms (e.g. brain for encephalus) has markedly improved the ability of subjects without a specific background to understand and make informed choices.

4 Possible Applications in Collective Cognition: Gathering Citizens' Opinion and Satisfaction

Gathering citizens' opinions and levels of satisfaction for services is vital for building a mature democracy and for the correctness of the relationship between citizens and the local government. While unpopular decisions have to be taken once in a while, disregarding too much citizens' judgment on political or social proposals, or being unable to identify the right 'sore spots' in handling the public goods can, in the long run, strain the relationship with citizens and deliver ineffectual government action. Since communication between citizens and local government (in this specific direction) is often done in writing, a correct and automated analysis of texts can help in delivering the communication itself to the right actors, and to proactively determine the general sentiment expressed by it. This is especially important as communication is often initiated by the citizen when something is going wrong, and as such the correct strength of the main feeling and the presence of other, more positive feelings that can counterbalance the main one have to be correctly assessed before an answer is given. We have devised a sentiment evaluation and analysis system, which will be briefly discussed here, that uses fuzzy linguistic textual analysis in order to help determining the main feeling expressed by a short-to-medium-length text (such as a letter, or more aptly nowadays a social network post), and the presence of eventual sub-feelings, and their intensity [12, 34].

4.1 Fuzzy Linguistic Textual Sentiment Analysis

Evaluating individual opinions are a difficult task because writing allows a gamut of expressions, limited only by the time at disposal and the linguistic proficiency of the writer.

There are several techniques, together called sentiment analysis, through which it is possible to find, in automatic way, the general sentiment (or the opinion) expressed within a text. In particular sentiment analysis methodologies allow to find the polarity of the text to which are applied. Given an input text, sentiment analysis methodologies provide as output its polarity, which could be thought as a class label. In fact, the problems of sentiment analysis are a specific instance of classification problems, approachable by using machine learning algorithms [4].

Sentiment analysis is employable in various application fields. In marketing, for example when a new product may be launched in different markets according to sentiment expression; the producers, analyzing the opinion of users, can forecast its popularity, and if it will be appreciated by the users and why. And more, sentiment analysis can be a specific support assets during elections; is possible to indirectly find if an elector prefer one candidate over another through the opinions that they express, and then align the opinions of a candidate accordingly. In our scenario citizens can express their opinion using comments or assess their feeling about contents published by an institution in writing.

The term opinion mining and sentiment analysis are often interchangeable and used with the same meaning but, as said in [28], "the concepts of emotions and opinions are not equivalent although they have a large intersection". It is clear that the emotional state of a person affect his opinion, but they are not the same things. A possible definition of the difference may lay in the fact that opinion mining methodologies are used to recognize the opinion polarity expressed in a text; sentiment analysis methodologies instead try to forecast the emotional state of the writer.

There are different technique in literature to approach sentiment analysis, each one with their peculiarity that tend to fit well within a particular problem from a specific domain. Applying a combination of fuzzy logic and text analysis is not a new approach for discovering the right sentiment in a text. A number of previous works in literature apply such techniques in tandem to obtain a more accurate result. For example, in [14] the authors present a system for the multi-domain sentiment analysis to take in account the different polarity of each term respect to domain in which it is. To do this they use fuzzy logic theory for modelling of the membership function to relate concepts and domains. Another related reference is [15] where the authors try to consider the affective state through affective space.

4.2   Our Approach to Fuzzy Text Analysis

When we discuss feelings we regard them mainly as fuzzy concepts, as each piece of content can be associated with different emotion, and such emotions can be expressed with different intensities, something which in the verbal discourse is usually represented using linguistic labels such as "a little", "a lot", "mildly", and son on. While in the psychology community does not exists a complete accord on which emotions are to be considered as basic (or fundamental, upon which other emotions are to be based), a number of sources have used Ekman and Friesen [17] classification of six basic emotions: happiness, anger, disgust, fear, sadness, surprise, or the alternative list from Parrott [32] that includes love, joy, surprise, anger, sadness, fear. Different lists exists, and even Ekman has updated his own classification including a number of other emotions [16], but the short list from [17] still seem to be one on which a vast literature on emotions is based, and even the popular press and media seem to have joined this bandwagon, as witnessed by a recent surge of popularity (see e.g. the box office champion Pixar movie "Inside

Out"). Such primary emotion are not necessarily mutually exclusive, and each text snippet can elicit a mix of them in different gradations, a situation that presents itself as a perfect match for the expressive power of Fuzzy Concepts. A reader can infact be "very happy" and "a bit surprised" or "quite angry" and "mildly sad" when commenting on a piece or a post, and such expression of feelings should be represented in full.

In order to pre-analyze the text that constitute our experimental corpus we use a classical approach that is aimed at quantifying the strength of each feeling as related to each snippet. A second phase, based on the Fuzzy inclusion of such expressions, is an interaction prediction phase that correlate the possible range of emotions with the previous interaction between two users. Information from the two phases can then be used for the computation of an appropriate feeling, represented by an emoji, in a hybrid fuzzy system. While the work carried on up to now is of a preliminary nature, good results have already been obtained by the use of probabilistic tools from the computational intelligence domain (in the general family of machine learning algorithms and stochastic processes).

A general schema of the process is as follows: we consider two actors per round: institution A, who writes the post or piece, and citizen B, who is reading the post and is willing to express an emotion about it. The method we devised starts with an analysis of the text, using the presence of specific terms that are annotated with specific sentiments, and searching for fuzzy quantifiers near the term in order to infer a degree of membership to the set of a specific emotion. This is a new twist on a classical text mining approach, and the experimental preliminary results have given a significative correlation with ground truth, which is gathered by having a third set of external reviewers grade the texts independently, using Likert-like scales. Recognising emotions in such texts is an easy task for humans, as reported by the reviewers, but the gradations tend to be masked by personal preferences or linguistic differences. The same process used to discover the feelings linked to the presumedly objective opinion expressed in a specific text (such as a press release, a facebook post or a series of twitters) are used in a following step in order to gather a prediction of the sentiment and its gradation expressed by the user. As a starting point we considered the Google Prediction Framework [19]. This framework allows to train a model with a training set, containing social network contents random gathered among public profiles or friend's profiles, previously manually labelled by users with one of the possible emotion labels. This procedure is generally accepted and provides good results without a long phase of training, but has a specific shortcoming to be usefully employed in our approach: as already mentioned before, for a human reader discovering the main emotion in a text is easy, but he can also recognize the others minor emotions that may be present in the text. This behaviour cannot be represented by standard models using training-set, but by implementing an additional phases it is possible to consider such aspect. After the training phase, the model can be used to predict the label of a new data, that have to be of the same type of the training set data. The output of the model consist in a probabilistic weight vector, where the position p_i is the probability that the data belong to the class p_i. This weight can be interpreted as fuzzy membership degrees of a certain feeling present in the original text. The constant updating in order to make predictions about text currents is carried out by the use of Linguistic Fuzzy Markov Chains. There are two main approaches relative to fuzzy Markov chains. The most extended one use a fuzzy relation over the cartesian product between the state space. In literature this approach are used e.g. in application for speech recognition and image classification [31, 20]. In a recent implementation of linguistic fuzzy Markov chains, Villacorta et al. [42] suggest to address the uncertainty of linguistic judgements by introducing fuzzy probabilities as an additional layer that is placed on top of Buckley's fuzzy Markov chains. This approach is the one we have followed for our implementation. In our experience the combination of Fuzzy classifiers for sentiment text prediction and fuzzy Markov chains for updating the system status brought a marked improvement in performance with respect to the standard way of employing Fuzzy classifiers by themselves, and this especially when communication is iterative and multi-party.

5 Conclusions and Further Work

In this paper we have discussed the idea that computational intelligence technologies, especially the ones deriving from fuzzy logic and soft computing, can be beneficial in order to improve aims and quality of citizen communication. We have listed a number of approaches that in our opinion will become useful in determining a number of aspects in such communications, and presented in a qualitative manner two approaches we have followed: one regarding the facilitation of comprehension in medical texts and diagnosis, the other facilitating the expression and classification of sentiment toward texts from an institution. Both example are from pilot studies, but already show that soft computing is useful in developing and deploying system that assist citizen's communication by giving to such systems flexibility, easy of use and a level of internal description that is much more near to the way humans plan and act in communication. A lot of work is still to be done in order to render such solutions commonplace: experimental confirmation of the existing methods has to be done in vivo in society, using real data from real communication problems for testing purposes; cardinality of the experiment has to reach real world level; techniques have to be affined, and parameters tuned in order to include case studies from different domains. But nonetheless we maintain that there is a prominent place for soft computing and fuzzy methods from computational intelligence in the future of smart cities.

References

1. Altunkaynak, A., Özger, M., Çakmakci, M.: Water consumption prediction of Istanbul city by using fuzzy logic approach. Water Resour. Manage. **19**(5), 641–654 (2005)
2. Ammar, S., Duncombe, W., Hou, Y., Wright, R.: Evaluating city financial management using fuzzy rule–based systems. Public Budgeting Finance **21**(4), 70–90 (2001)

3. Becker, H.: Computing with words and machine learning in medical diagnostics. Inf. Sci. **134** (1–4), 53–69 (2001)
4. Buche, A., Chandak, M.B., Zadgaonkar, A.: Opinion mining and analysis: a survey. Int. J. Nat. Lan. Comput. **2**(3), 39–48 (2013)
5. Chen, B., Cheng, H.H.: A review of the applications of agent technology in traffic and transportation systems. IEEE Trans. Intell. Transp. Syst. **11**(2), 485–497 (2010)
6. Chomsky, N.: Three models for the description of language. IRE Trans. Inf. Theory **2**(3), 113–124 (1956)
7. Chomsky, N.: On certain formal properties of grammars. Inf. Control **2**(2), 137–167 (1959)
8. Chomsky, N.: Syntactic Structures. Walter de Gruyter, Berlin (2002)
9. Consumer Health Vocabulary Initiative: Open Access Collaboratory Consumer Health Vocabulary http://consumerhealthvocab.org. Accessed 10 Nov 2015
10. Cosido, O., Loucera, C., Iglesias, A.: Automatic calculation of bicycle routes by combining meta-heuristics and GIS techniques within the framework of smart cities. In: 2013 International Conference on New Concepts in Smart Cities: Fostering Public and Private Alliances (SmartMILE), Gijon, 11–13 Dec 2013
11. D'Aleo, F., D'Asaro, F.A., Perticone, V., Rizzo, G., Tabacchi, M.E.: Agents displacement in arbitrary geometrical spaces: an evolutionary computation based approach. In: Loiseau, S., Filipe, J., Duval, B., Van den Herik, J. (eds) Proceedings of the International Conference on Agents and Artificial Intelligence, Lisbon, 10–12 Jan 2015
12. D'Aleo, F., Perticone, V., Rizzo, G., Tabacchi, M.E.: Can you feel it will you tell me. Encouraging sentiment expression on the web. In: Airenti, G., Bara, B.G., Sandini, G. (eds) Proceedings of the EuroAsianPacific Joint Conference on Cognitive Science, Turin, 25–27 Sept 2015
13. D'Asaro, F.A., Perticone, V., Tabacchi, M.E.: A fuzzy methodology to alleviate information overload in eLearning. In: Pasi, G., Montero, J., Ciucci, D. (eds) Proceedings of the 8th conference of the European Society for Fuzzy Logic and Technology (EUSFLAT-13), Milan, 11–13 Sept 2013
14. Dragoni, M., Tettamanzi, A.G., Da Costa Pereira, C.: A fuzzy system for concept-level sentiment analysis. In: Presutti, V. et al (eds) Semantic Web Evaluation Challenge, Anissaras, 25–29 May 2014
15. Dzogang, F., Lesot, M.J., Rifqi, M., Bouchon-Meunier, B.: Expressions of graduality for sentiments analysis—a survey. In: 2010 IEEE International Conference on Fuzzy Systems (FUZZ), Barcelona, 18–23 July 2010
16. Ekman, P.: Facial expressions. In: Dalgleish, T., Power, M.J. (eds) Handbook of Cognition and Emotion, John Wiley & Sons Ltd, Chichester (1999)
17. Ekman, P., Friesen, W.V.: Constants across cultures in the face and emotion. J. Pers. Soc. Psychol. **17**(2), 124–129 (1971)
18. Formica, A.: Concept similarity in fuzzy formal concept analysis for semantic web. Int. J. Uncertainty Fuzziness Knowl. Based Syst. **18**(02), 153–167 (2010)
19. Google: Google Prediction API. https://cloud.google.com/prediction/. Accessed 10 Nov 2015
20. Hatt, M., Lamare, F., Boussion, N., et al.: Fuzzy hidden Markov chains segmentation for volume determination and quantitation in PET. Phys. Med. Biol. **52**(12), 3467–3491 (2007)
21. Herrera, F., Alonso, S., Chiclana, F., Herrera-Viedma, E.: Computing with words in decision making: foundations, trends and prospects. Fuzzy Optim. Decis. Making **8**(4), 337–364 (2009)
22. Juan, Y.K., Wang, L., Wang, J., Leckie, J.O., Li, K.M.: A decision-support system for smarter city planning and management. IBM J. Res. Dev. **55**(1.2), 3–1 (2011)
23. Kaltenrieder, P., Portmann, E., D'Onofrio, S.: Enhancing multidirectional communication for cognitive cities. In: 2nd International Conference on eDemocracy & eGovernment, pp. 38–43. IEEE, Quito (2015). doi:10.1109/ICEDEG.2015.7114476
24. Kaltenrieder, P., Portmann, E., Binggeli, N.K., Myrach, T.: A conceptual model to combine creativity techniques with fuzzy cognitive maps for enhanced knowledge management. In: Fathi, M. (ed.) Integrated Systems: Innovations and Applications, pp. 131–146,. Springer (2015). doi:10.1007/978-3-319-15898-3_8

25. Keselman, A., Slaughter, L., Arnott-Smith, C., et al.: Towards consumer-friendly PHRs: patients' experience with reviewing their health records. In: AMIA Annual Symposium Proceedings, Chicago, 10–14 November 2007
26. Keselman, A., Smith, C.A., Divita, G., et al.: Consumer health concepts that do not map to the UMLS: where do they fit? J. Am. Med. Inform. Assoc. 15(4), 496–505 (2008)
27. Lee, S.J., Back, A.L., Block, S.D.: Stewart SK (2002) Enhancing physician-patient communication. ASH Education Program Book 1, 464–483 (2002)
28. Liu, B.: Sentiment analysis and subjectivity. In: Indurkhya, N., Damerau, F.J. (eds.) Handbook of Natural Language Processing, 2nd edn, pp. 627–666. Chapman & Hall/CRC, London (2010)
29. Martínez, L., Ruan, D., Herrera, F.: Computing with words in decision support systems: an overview on models and applications. Int. J. Comput. Intell. Syst. 3(4), 382–395 (2010)
30. Nha, V.T.N., Djahel, S., Murphy, J.: A comparative study of vehicles' routing algorithms for route planning in smart cities. In: 2012 First International Workshop on Vehicular Traffic Management for Smart Cities, Dublin, 20 Nov 2012
31. Patel, I., Rao, Y.S.: Technologies automated speech recognition approach to finger spelling. In: 2010 International Conference on Computing Communication and Networking Technologies (ICCCNT), 1–6 July 2012
32. Parrott, W.G.: Emotions in social psychology: essential readings. Psychology Press, New York (2001)
33. Parry, D.: Evaluation of a fuzzy ontology-based medical information system. Int. J. Healthc. Inf. Syst. Inf. 1(1), 40–51 (2006)
34. Perticone, V., D'Aleo, F., Rizzo, G., Tabacchi, M.E.: Towards a fuzzy-linguistic based social network sentiment-expression system. In: Alonso, J., Bustince, H., Reformat, M. (eds) Proceedings of the 2015 Conference of the International Fuzzy Systems Association and the European Society for Fuzzy Logic and Technology, Gijon, 30 June–3 July 2015
35. Plain Language Action and Information Network: Federal Plain Language Guidelines. http://www.plainlanguage.gov/howto/guidelines/FederalPLGuidelines/FederalPLGuidelines.pdf. Accessed 10 Nov 2015 (2011)
36. Portmann, E., Kaltenrieder, P., Zurlinden, N.M.: Applying fuzzy ontologies to implement the social semantic web. ACM SIGWEB Newsl. ACM 10(1145/2682914), 2682918 (2014)
37. Seising, R., Tabacchi, M.E.: A very brief history of soft computing. In: Pedrycz, W., Reformat, M.Z. (eds) Proceedings of the 2013 Joint IFSA World Congress, NAFIPS Annual Meeting (IFSA/NAFIPS), Edmonton, 24–28 June 2013
38. Trillas, E., Termini, S., Tabacchi, M.E., Seising, R.: Fuzziness, cognition and cybernetics: an outlook on future. In: Alonso, J., Bustince, H., Reformat, M. (eds) Proceedings of the 2015 Conference of the International Fuzzy Systems Association and the European Society for Fuzzy Logic and Technology, Gijon, 30 June–3 July 2015
39. Truong, H.B., Nguyen, N.T., Nguyen, P.K.: Fuzzy ontology building and integration for fuzzy inference systems in weather forecast domain. Intelligent Information and Database Systems, pp. 517–527. Springer, Berlin Heidelberg (2011)
40. Turing, A.M.: On computable numbers, with an application to the Entscheidungsproblem. J. Math. 42(1), 230–265 (1936)
41. Unified Medical Language System: U.S. National Library of Medicine Bethesda. https://www.nlm.nih.gov/research/umls/. Accessed 10 Nov 2015
42. Villacorta, P., Verdegay, J., Pelta, D.: Towards fuzzy linguistic Markov chains. In: Pasi, G., Montero, J., Ciucci, D. (eds) Proceedings of the 8th conference of the European Society for Fuzzy Logic and Technology (EUSFLAT-13), Milan, 11–13 Sept 2013
43. Wikimedia Foundation: Wikipedia:Simple English Wikipedia. https://simple.wikipedia.org/wiki/Wikipedia:Simple_English_Wikipedia (2011). Accessed 10 Nov 2015
44. Winter, J.A.: Doctor, can we talk? Physician-patient communication issues that could jeopardize patient trust in the physician. S. D. J. Med. 53(7), 273–276 (2000)

45. Yadav, J., Kharat, V., Deshpande, A.: Fuzzy description of air quality using fuzzy inference system with degree of match via computing with words: a case study. Air Qual. Atmos. Health **7**(3), 325–334 (2014)
46. Zadeh, L.A.: Fuzzy logic = computing with words. IEEE Trans. Fuzzy Syst. **4**(2), 103–111 (1996)
47. Zadeh, L.A.: Applied soft computing–foreword. Appl. Soft Comput. **1**(1), 1–2 (2001)
48. Zielstorff, R.D.: Controlled vocabularies for consumer health. J. Biomed. Inform. **36**(4), 326–333 (2003)

Digital Personal Assistant for Cognitive Cities: A Paper Prototype

Patrick Kaltenrieder, Elpiniki Papageorgiou and Edy Portmann

Abstract This chapter presents an evaluation and initial testing of a meta-application (meta-app) for enhanced communication and improved interaction (e.g., appointment scheduling) between stakeholders (e.g., citizens) in cognitive cities. The underlying theoretical models as well as the paper prototype are presented to ensure the comprehensibility of the user interface. This paper prototype of the meta-app was evaluated through interviews with various experts in different fields (e.g., a strategic consultant, a small and medium-sized enterprises cofounder in the field of online marketing, an IT project leader, and an innovation manager). The results and implications of the evaluation show that the idea behind this meta-app has the potential to improve the living standards of citizens and to lead to a next step in the realization and maturity of the meta-app. The meta-app helps citizens more effectively manage their time and organize their personal schedules and thus allows them to have more leisure time and take full advantage of it to ensure a good work-life balance to enable them to be the most efficient and productive during their working time.

1 Introduction

Optimization and efficiency are keywords in today's economy. This trend is once again reducing the separation of work and leisure time. Before the industrial revolution as well as in ancient cultures other than those of the western world, there

P. Kaltenrieder (✉) · E. Portmann
Institute of Information Systems, University of Bern, Bern, Switzerland
e-mail: patrick.kaltenrieder@iwi.unibe.ch

E. Portmann
e-mail: edy.portmann@iwi.unibe.ch

E. Papageorgiou
Department of Computer Engineering, Technological Education Institute
of Central Greece, Lamia, Greece
e-mail: epapageorgiou@mail.teiste.gr

© Springer International Publishing Switzerland 2016
E. Portmann and M. Finger (eds.), *Towards Cognitive Cities*,
Studies in Systems, Decision and Control 63, DOI 10.1007/978-3-319-33798-2_6

was no conscious separation of work and leisure time [45]. Economic development reduced the number of hours spent at work over time, thus increasing free time [2]. This strict separation and the growing amount of leisure time started to diminish at the end of the 20th and beginning of the 21th century. The separation of work and leisure has become ambiguous for not only upper-level managers but also middle and low-level managers as well as, more and more, common employees [23]. This leads to a situation wherein employees are accessible to companies nearly 24/7. Thus, most social and technological changes have led to new attitudes toward and ideas of work. Electronic markets are location as well as time independent. Society calls for new working conditions that make independence and a reasonable work-life balance possible [26].

In addition, there has also been a change in the human mindset. Individuals are not employees of a single company anymore; they are starting to become their own innovation managers because they have to constantly define and re-define themselves, and their unique selling proposition, against a constantly evolving work environment to remain competitive. Individuals are simultaneously working in different fields and work both as employees for companies and as employers as they start to be entrepreneurs [26]. Methods have been developed to allow employees to create personal business models to adapt to changing conditions in the market and to create satisfying work opportunities [5]. This also includes the organization of personal schedules. Leisure time has to be included in this planning to ensure a certain amount of separation between work and leisure and to remain capable of the constant re-invention of oneself. Changing work environments can produce negative effects on individual workers, such as huge workloads and pressures to meet the high expectations of demanding and growing companies, and on society such as rising unemployment rates. As a result of these developments in the work world, Ruh [39] perceives it to be necessary to adapt our personal schedules. Sufficient leisure time has to be included to ensure periods of relaxation, recovery and compensation [39]. To allow one to optimize their personal schedule and be as efficient as possible in all aspects and situations of life, we propose a meta-app named *cogniticity*. This meta-app has the ability to improve people's personal schedules as well as cooperation between various stakeholders of a cognitive city.

Such applications can allow individuals to develop into knowledgeable workers [26] that are supported and enhanced by computers. The use of information systems introduces a higher degree of abstraction, independence and responsibility into work [26]. Travel tools that help users optimize their time use can make a big step into this direction.

Various applications have already been developed to support citizens (e.g., Waze,[1] Snips,[2] Google Now,[3] and IBM Watson[4]) in addition to apps that provide

---

[1] http://www.waze.com.

[2] http://snips.net/.

[3] http://www.google.com/landing/now/.

[4] www.ibm.com/smarterplanet/us/en/ibmwatson.

advice and recommendations about travel and places. However, tools enabling semi-automated decision making have yet to be developed. Such semi-automated decision making tools intend to go one step further and add a new dimension to existing travel apps by including a decision support system that automatically changes routes, travel plans, and meeting places. This is the objective of the meta-app proposed in this chapter. By facilitating commuting, this meta-app can help people optimize their time scheduling and thus lead to a better life-work balance.

Such applications can lead to cognitive cities a smart city enhanced with cognitive computing, to address the challenges complex world challenges (cf. Sect. 2.1). At least part of which was first introduced in Mostashari et al. [30]. Thus, this chapter complements existing research and presents a tool that can bring the realization of a cognitive city one step closer as this meta-app serves as a bridging element between the theoretical and conceptual implementation of cognitive cities and the real world. The meta-app is developed respecting fuzzy logic. Because traditional logic is inadequate to account for the uncertainty and imprecision of human reasoning and its environment (a concept of paraconsistent logic, as fuzzy logic) is needed. Fuzzy cognitive maps (FCMs) are an application of fuzzy logic and form the bases of our meta-app. Cognitive computing allows one to handle today's large and complex datasets, which cannot be analyzed by traditional computing techniques alone [13]. As a digital personal assistant (i.e., the meta-app) in the Semantic Web, the meta-app improves the daily life of a citizen and enables the implementation of Cognitive Computing through real-time assistance (e.g., by providing routes, by providing possibilities, and supporting decision making) by simultaneously drawing information from several resources (cf. Sect. 2.3).

The next section presents the theoretical background of the meta-app used in this chapter. The meta-app and its benefits as well as challenges faced by the users and stakeholders of a cognitive city are introduced in the third subchapter. The fourth section presents the main focus of the chapter, namely, an evaluation and its result and implications. The meta-app in the smart and—on this basis—cognitive city context are embedded in the fifth subchapter. The chapter is concluded with an outlook on future research.

2 Background

The intended meta-app is based on the theories of fuzzy sets, fuzzy logic, FCMs and cognitive computing to apply soft computing methods for cognitive computing in large urban systems. Cognitive computing [29] and computational thinking [50] enhanced by an intelligence amplification loop [7] form the backbone of the meta-app, thereby improving the interactions of stakeholders (e.g., users) and systems. FCMs [18, 34] are incorporated in the meta-app to aggregate the knowledge of the involved stakeholders, to represent this knowledge through its graphical modeling capabilities and to enable reasoning based on the aggregated knowledge. The following paragraphs define the concept of smart and cognitive cities, followed by an overview on the above-mentioned concepts and their relevance to the meta-app.

2.1 Smart and Cognitive Cities

As the increasing urban population [49] makes it necessary to have functioning systems to ensure a well-working living together among citizens [20], the concepts of smart cities show new important ways of city management. Investments in human and social capital and in communication infrastructure can enable sustainable growth and a high living standard for citizens and thus the realization of a smart city [4]. The smartness of a smart city manifests itself in the smartness of its economy, people, governance, mobility, environment and living [11]. Kaltenrieder et al. [15, 17] perceive smart cities as based on technology developed to constantly enhance their interaction with citizens by improving cooperation between humans and machines.

Cognitive cities are an enhancement of smart cities based on the addition of cognition and cognitive systems. A cognitive system is able to take advantage of the past to learn (i.e., by machine learning) and such a system recognizes and adapts to changes [30]. The concept of cognitive cities, as defined by Kaltenrieder et al. [15, 17], utilizes connectivism [42] to enhance smart cities. Connectivism is a type of cognition theory based on the fact that knowledge bases increase steadily. This makes it impossible for humans to learn only with the help of their own experiences and perceptions. Instead, they have to consider the experiences of others (i.e., networks are very important) and increasingly rely on the assistance of computers to be able to address the scale and complexity of information [41, 42]. Thus, for a city to address today's challenges and increasing numbers of citizens and to ensure the proper functioning of its systems, it has to enable the cooperation of humans and machines, to take full advantage of the ever increasing knowledge bases. As observed in the introduction, an important aspect of citizens' lives is the work-life balance (i.e., people need sufficient leisure time to be able to recuperate from work and to be as efficient as possible while working). Tools that optimize people's time scheduling might thus enhance an important aspect of citizens' daily lifes and improve their living environment.

2.2 Fuzzy Cognitive Maps

This section of the background discussion provides a short overview of fuzzy logic and FCMs.

Because human reasoning and its environment are characterized by imprecision and uncertainty, we require a concept of logic that is able to consider these phenomena into account. Traditional bivalent logic only includes two truth values, *true* and *false*, and is thus not alone able to accomplish this task. To overcome this shortcoming, Zadeh [52] developed the concept of fuzzy logic, which is based on fuzzy set theory [51]. In traditional set theory, a set is defined by its elements, and an element either belongs to a set or it does not. In fuzzy set theory, an element is only included in a set to a certain degree, and thus, it has different grades of membership to several fuzzy sets. From this follows that a fuzzy set is defined by a membership function f(X), which takes on values between 0 and 1. The closer the value is to 1, the higher is the element's grade of membership to the fuzzy set [51]. Linguistic variables can be used as labels for the membership degree, like *weak*, *high* etc. [52]. Thus, fuzzy logic augments bivalent truth by partial truth, where a proposition can be anything between false or true (i.e., it can take on any degree of truth [52]).

The concept of FCMs [21] is an extension of traditional cognitive maps [46] based on fuzzy logic. An FCM consists of nodes, which are concepts (e.g., outcomes or goals) and edges which connect the nodes with each other and describe the causal relationships between them. The edges can be represented as an adjacency matrix. The element m_{kl} in matrix M denotes the weight of the edge that connects the two nodes C_k and C_l and takes values between -1 and 1. A positive element m_{kl} denotes a positive causal effect from C_k on C_l and a negative element a negative effect. The element is equal to zero if there is no effect. As the causal effect from C_l on C_k can be different than the effect from C_k on C_l (i.e., m_{kl} does not equal m_{lk}), the matrix M does not have to be symmetric [10].

As the concepts of an FCM influence each other, a FCM changes over time. FCMs that have the same concepts, have matrices with identical dimensions and can thus be aggregated [10, 18, 22].

In a further step, FCMs can be seen as useful representations of individual and collective mental models that can be transformed in order to achieve a computing with words (CW) process without loss of information [35]. CW is a new concept and is used at this point as natural language concept of cognitive computing. It is in essence a methodology for reasoning, computing and decision making with information described in words and sentences. Accordingly, this information, described in natural language, is imprecise [27].

2.3 Cognitive Computing

This section of the background provides a brief overview of cognitive computing, which is an important basis of our meta-app. Fields of cognitive computing are connectivism, which was already introduced in Sect. 2.1, computational thinking, and the intelligence amplification loop.

To enable interaction between humans and machines, and thus enable cognitive cities, cognitive computing is needed, because traditional systems are unable to handle the scale, uncertainty and complexity of data [13]. Because cognitive systems are able to learn and process large amounts of complex data, it will be possible to address "*[...] the world as it really is: emergent, not deterministic*" and thus to handle "*[...] its complexity and unpredictability*" [20, pp. 128, 129]. A cognitive computing system (automatically) learns based on data and experience by detecting patterns and deriving meaningful information from texts, images and other sources. Artificial intelligence and machine learning is considered to be the foundation of cognitive computing. The goal of artificial intelligence is to represent data in a way such that the data can be manipulated and that people can make inferences based on such data. Because many data types are unstructured, ambiguous and uncertain, methods must be developed to enable computers to handle these data. It is assumed that cognitive computing can enable computers to mimic human thinking [13] by developing "*a coherent, unified, universal mechanism inspired by the mind's capabilities*" [29]. This can lead to new learning, computing and programming systems and applications that can simultaneously take advantage of data from many different sources [29].

One aspect of cognitive computing is also computational thinking [50]. Its goal is to find solutions to complex problems and to understand human behavior with the help of computer sciences. By reducing, transforming, or decomposing data, complex problems can be solved. Thus, computational thinking allows one to operate on different levels of abstraction and to mechanize these levels through precise notations and models [50].

Cognitive computing enables improved reasoning. Engelbart [7] previously proposed a mechanism to enhance human reasoning in 1962 through the assistance of computers. This idea is captured by the concept of the intelligence amplification loop, which proposes that both humans and computers learn through mutual interaction [19, 37]. This process is characterized by emergence (i.e., the phenomena whereby new properties emerge unplanned from the interaction of a system's components) [9]. This is an important concept in the development of the meta-app because cooperation among a cognitive city's stakeholders increases intelligence bottom-up (i.e., grass-roots growth) through the presented intelligence amplification loop.

2.4 Stakeholder Analysis in Cognitive Cities

The meta-app developed in this paper can be used by different stakeholders of a cognitive city. Relevant stakeholders of a cognitive city include citizens, government representatives, and service providers. Because the interests and preferences of different stakeholders are not identical or even well defined, city governance faces various challenges. It is difficult to find optimal solutions, and compromises among the stakeholders have to be reached [30].

The meta-app allows the input of the preferences and interests of its users. This knowledge base makes it possible to enhance the meta-app, and thus the cooperation between different stakeholders, and can therefore facilitate urban governance.

Table 1 Stakeholders of the meta-app

Group	Stakeholder	Specifications
Personal	User	Citizens
Governmental	Governmental services	E.g., transportation services, communal services
Business	Third-party services	E.g., restaurants, taxi companies, healthcare providers
Business	Third-party providers	E.g., providers of existing apps, developers of apps

In order for the meta-app to be successful, its various stakeholders have to be considered. The following is a list of the various stakeholders of the app. This list is not comprehensive as other stakeholders may appear in the future. An initial list of the stakeholders affected by the meta-app was generated by Kaltenrieder et al. [17]. This list was refined as the meta-app was developed further, and an initial prototype was sketched [16]. The stakeholders can still be grouped into three main groups (i.e., personal, governmental and business). These groups represent the sources of the information. Table 1 shows the various stakeholders of the meta-app applied in the evaluation.

3 The Meta-App

As explained in the introduction of the chapter, the main focus of this chapter is the evaluation of the meta-app. The meta-app and its history have to be explained and understood to be able to conduct an evaluation. This section gives an overview of the history of the meta-app.

Interaction and communication in a cognitive city is a key resource for all involved stakeholders. Researchers have proposed the use of FCMs and computational intelligence techniques to enhance stakeholder management [36]. They developed a tool for stakeholder mapping that contains searchable and browsable knowledge structures of different granularity. Existing applications already support citizens of cognitive cities in different situations (e.g., Waze, Snips, Google Now, and IBM Watson). Various tools have also been developed to support travelling and commuting by providing advice and recommendations on travel routes, places and means of transportation. However, tools that facilitate semi-automated decision making have yet to be developed. The meta-app *cogniticity* applies knowledge aggregation, representation and reasoning to smart and cognitive cities. The meta-app enhances and supports semi-automated decision making in a cognitive city. By facilitating commuting, this meta-app can thus help people optimize their time scheduling and thus lead to a better work-life balance. The architecture of the meta-app was developed by Kaltenrieder et al. [17]. Focusing on the fuzzy analytical hierarchy process, the theoretical decision making logic of the meta-app was developed by Kaltenrieder et al. [15]. Currently, the stage of development of the meta-app is the paper prototyping phase [16]. This allowed us to conduct evaluations with several experts to further refine the meta-app (cf. Chap. 4).

4 Evaluation

The main focus of this chapter is the performance of an evaluation. A current version of a paper-based prototype [16]. Snyder [48] is the foundation of the evaluation. The evaluation was performed with several experts. Due to their various backgrounds (i.e., the expert panel includes a strategic consultant, an IT project leader, a small and medium-sized enterprises (SME) cofounder in the domain of online marketing, and an innovation manager), these experts possess a wide range of knowledge. This range of knowledge permits us to gain a broad view of the different aspects of the paper prototype from various perspectives (e.g., user, third-party provider and data source). The goals of the evaluation were to gather expert feedback to improve the current paper prototype and to initialize a further, more advanced version of the paper prototype. The results of the prototype are used to improve the requirement identification and analysis as well as to enable a step toward a first version of a software prototype.

4.1 Paper Prototyping

Prototypes allow the simulation of a system's or product's most important functions through user testing [8, 40]. There exist several forms of prototypes, from rough drafts on paper to "lo-fi" (i.e., low fidelity) mockups using stripped-down interface designs to "click-through" prototypes. Step by step, these prototypes represent an increasingly more realistic system and thus create the illusion of a finished product [8]. The goal is to develop a prototype that is simple to handle and easy to understand without using a lot of time and effort [40]. Prototyping allows one to visualize and test an application before it is implemented and serves as a tool of inquiry. It helps to take a designing attitude, which means that one takes into consideration ideas that are only in their first step of development, but also helps to discard them rapidly if they do not seem to be fruitful. This also makes one examining several possibilities and accepting uncertainty that is prevalent until the direction of the design gets clearer, which can be very valuable in the development of applications [33].

Paper prototypes are a special form of prototypes that are mainly hand drawn [40] and can be applied as an interactive medium that links to digital information and services [43]. The use of prototypes allows to explore, communicate and evaluate early interface designs [1] and to invite people with little-to-no technical background to be a part of the design process [25].

Snyder [48] proposed the following definition of paper prototyping: "*Paper prototyping is a variation of usability testing where representative users perform realistic tasks by interacting with a paper version of the interface that is manip- ulation by a person 'playing computer,' who doesn't explain how the interface is intended to work.*"

The use of the paper prototype helps us follow a design science research approach [14] in the development process of the meta-app. The advantages of paper prototypes are, amongst others, that design ideas can be rapidly externalized with minimal costs and numerous alternatives can be generated and tested early on in the design cycle because of the iterations by designers. Emergent issues can be resolved as early as possible and process related stakeholders (i.e., users and other designers) are more willing to give true feedback of the design since it appears rough and informal. The process of giving feedback, often in terms of thinking aloud, raises the motivation to explore numerous different ideas [1, 24, 48].

Adversely, paper prototyping needs a facilitator in the majority of cases. A facilitator is someone who comprehensively understands the application and can adequately demonstrate the functionality to the test participant. In other words, the test participant has to be instructed by the facilitator. The facilitator changes paper screens and explains how the various elements of the design would work, while the test participant has the possibility to give feedback during the simulation sequence. Without human intervention (i.e., from the facilitator), the test participant would not understand how the system should respond. Thus, a full interaction is limited, because there is still a certain dependency between the test participant and the facilitator [3, 6]. Furthermore, it is difficult to demonstrate certain concepts, such as loading time of a screen, with paper [25].

Paper prototyping is a promising method to visualize a system that allows testing it with groups of users to obtain their feedback for a further iteration of the development of the system [1].

We use Microsoft PowerPoint as a mean for the evaluation as it has the same layout as a printed handout [43]. The evaluation was conducted by means of a walkthrough, where the test users (i.e., the experts) should ask questions and give feedback. Prototype walkthroughs allow learning about design [12].

4.2 Setting

The setting and performance of the evaluation are based on Snyder [48]. The evaluation was conducted using a Microsoft PowerPoint presentation following the guidelines presented in Sect. 4.1. The evaluation was separately conducted with five different experts. The attendees of the evaluation were the test participant as user of the meta-app (i.e., expert), one researcher acted as the moderator/computer, and one researcher acted as an observer who took notes during the interviews. The evaluation was conducted between 24th May 2015 and 11th June 2015 in Switzerland, Germany and Greece. Because the evaluation was conducted in three countries, two researchers conducted the interviews using interview guidelines [cf. 28]. All interviews were taped for a sound analysis, and notes were taken according to the utilized protocol. The evaluation was scheduled to last approximately 60 min (i.e., including the briefing and debriefing of the participant). A version of the conducted evaluation is available on our blog through the following link:

http://smartandcognitivecities.blogspot.ch/.

The participating experts possess a wide range of knowledge (e.g., knowledge from a strategic consultant, an IT project leader, an SME co-founder, and an innovation manager). This range of knowledge permits a sound view and field knowledge of the different aspects and perspectives (e.g., users, third-party providers, and data sources) of the paper prototype to be obtained. The only perspective not included in the evaluation was the city perspective because we primarily wanted to focus on the user and third-party perspective (i.e., bottom-up development of smart and cognitive cities [47]). The cognitive city perspective will be a priority in the next phase of prototyping when a newer version (i.e., mock-up) will be tested. In the next subchapter, the performance of the evaluation will be explained in more detail.

4.3   The Performance of the Evaluation

After the briefing (i.e., introducing the idea, motivation, goals of the paper prototype and explaining the rules), the evaluation was conducted in three phases. In the first phase, the user was introduced to the meta-app *cogniticity*, and the available functions were explained. Additionally, the user learned from which sources the data would be gathered and to which degree such data would be used (i.e., which data are needed to run the meta-app). Figure 1 shows the icon of the meta-app and exemplary data sources of the meta-app.

In the second phase the user was presented with the setting-up of a user account and its basic functions. First, the registration of a profile was presented, and the process of a new registration was detailed. Additionally, logging into the meta-app is possible via synchronization with an existing profile of another social networks site (e.g., Facebook, and Google+). Second, the format settings were introduced using the date format as an illustrative example. Third, the setting for the decision making mode had to be chosen. In the menu *decision making,* the user can determine to which degree the meta-app should support his decision making (i.e., the option *manual* means no support at all, *semi-automatic* means to a certain degree, and *automatic* means completely supported decision making). Figure 2 shows the decision making screens.

Further, the user can indicate their preferences and rank these preferences. The options *travel* and *food* are already provided, but the user can also add further preferences. The ranks of the preferences will influence the suggestions provided by the app to the advantage of the user and facilitate decision making. The preferences were explained with the examples of *travel* and *food.* In the case of travel, two criteria were already implemented: *price* and *time*. Additionally, there is the possibility to add additional criteria. To illustrate this functionality, the criterion

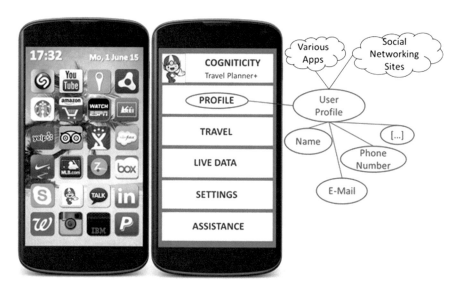

Fig. 1 Icon and exemplary data sources of the meta-app

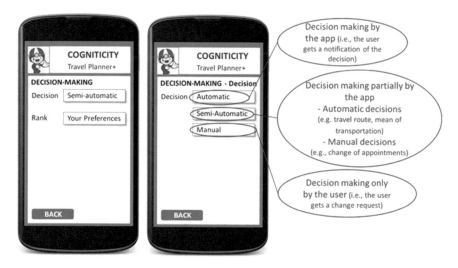

Fig. 2 Decision making screens of the meta-app

convenience was added. For every criterion, the user can indicate their importance on a slider ranging from *very important* to *not important at all* (see Fig. 3).

The same procedure was applied for the preference *food*. In this case, the user had to indicate which type of food (e.g., *Italian* or *Swiss*) that they prefer, by again indicating how important a range of criteria are to them (i.e., this time, the criteria were *Asian*, *Italian*, *Mexican*, and *Swiss*).

Fig. 3 Decision making preferences of the meta-app

Following the presentation of the *decision making* setting, the *privacy* menu was introduced. This functionality allows the user to decide which data will be shared only with the system (i.e., the data must be shared with the system because otherwise the meta-app would not work) without allowing other users to access it and which data can be shared in an anonymized manner with other users (which would allow the meta-app to be enhanced). Figure 4 shows the privacy screen of the meta-app.

Finally, the *synchronization* menu was presented to the user. This menu allows the user to determine with which sites (i.e., social networking sites, such as Facebook or Google+) they want to be synchronized (i.e., allowing the transmission of the data from her/his other profiles to the meta-app). Of course, the synchronization can only occur if the user has a valid login to the social networking sites.

In the final phase, an example of the functionality of the app was presented to the user. The example, presented as a use case, was based on various examples found in the literature [15–17].

This section shows a use case that explains the functionality of the meta-app from a user's perspective.

> It is Monday, the 1st of June 2015. The user has a whole-day meeting with his clients. In the evening, he has an appointment with friends at a restaurant. The following chronology shows how the meta-app can assist the user by providing important information and even recalculating the travel schedule and choosing a new location after the chosen one could not be reached.
>
> **17:33** A notification appears to remind the user that the tram will leave the station «Kappeli» in 10 minutes at 17:43. The user closes the notification.

Fig. 4 Privacy screen of the meta-app

17:38 A notification appears to remind the user that the tram will leave the station «Kappeli» in 5 minutes at 17:43. The user opens the meta-app by clicking on the *cogniticity* logo. The fixed appointment appears with all of the needed information (e.g., date, time, location and travel). He clicks on the function *travel* to change the means of public transportation from tram to taxi. After submitting the modification, a new notification appears: "The taxi will wait in front of the office in 30 minutes at 18:08." The user closes the notification (see Fig. 5).

17:53 A notification appears to remind the user that the taxi will wait in front of the office in 15 minutes at 18:08. The user closes the notification.

18:03 A notification appears to remind the user that the taxi will wait in front of the office in 5 minutes at 18:08. The user closes the notification.

18:09 The meeting goes on and ends too late to reach the taxi. A new notification appears: "The appointment could not be reached therefore a new location for the dinner was selected. New location: «Peking Garden», Altstetten. The bus 89 will leave the station «Kappeli» in 14 minutes at 18:23." (see Fig. 6).

18:23 The user takes the bus and goes to the new selected location to meet his friends.

The meta-app can improve the user's standard of living by acting as a digital personal assistant. The ability to share data with other users allows the app to choose another location so that everyone can be on time at the same location. It is also important to note that the app primarily acts on the events of the user who has installed the meta-app and triggered the meeting. The events of the invited people (i.e., also using the meta-app) are also mentioned as a secondary aspect.

4.4   Results and Implications

This section presents the results and implications of the evaluation. The feedback provided by the experts was analyzed and anonymized for the presentation in the

Fig. 5 Change of means of transportation

chapter. For comprehensibility, the results will be discussed in the sequence in which they were given during the evaluation.

The experts had different angles to approach the paper prototype, starting with the idea of the meta-app through the conceptual and technological viewpoint to the user perspective. Briefly, the idea was highly regarded by the experts. The underlying complexity of an actual implementation is high. The understandability of the presented concepts was high according to the experts.

Fig. 6 Notification about a
new location

The presented overview of the functions demonstrated the complexity of conducting an evaluation with a paper prototype as the experts started to ask questions and provide feedback on aspects occurring later in the interview. The focus on the basic functions started with the home screen of the app, where there is currently (in the version used in the evaluation) no initial login via username and password. The addition of a username and password request to start the meta-app was suggested by an expert. Second, the usability and complexity based on entering a name (i.e., first name, last name, together or separated in two fields) was also suggested by an expert. Finally, an expert indicated that a question about gender is missing. This request concerns usability as well as matching and synchronization. We were suggested to further analyze the registration process and to question what type of information is really needed.

One expert mentioned that it would be necessary to provide certain information, such as the current location of oneself and of friends, in advance. Such information would be more important than certain current settings. Without this information the localization process would not work.

The same expert indicated that users should be able to choose if they want to always go to the same restaurant or have some variety despite the selected preferences. Additionally, the function of specifying the preferences was highly regarded. Nevertheless, we were advised to consider other alternatives to avoid overlapping between preferences (e.g., introduce a rank), which would lead to dissatisfaction.

Overall, the meta-app appears as easily operated and clearly arranged. One expert recommends to elaborate the design and to carefully choose the color scheme because it influences usability and thus the satisfaction of users. The user interface should also be intuitive.

Various experts mentioned the importance of a possible specification of the decision making (i.e., which decisions are manual and which decisions are automated). This would prevent the user from a perceived loss of control.

One thing that is missing and of importance, according to one expert, is a decision aid. Sometimes, it is confusing as to which data concerning decision making should be given or shared: therefore, a decision aid would help users understand what type of information is needed by the meta-app and to which degree such information will be shared with the system and with other users. Instructions in general are requested from all experts to enhance usability.

Another expert stated that the focus should be on the creation of a cross-platform accessible API that users can use in all of their devices because people mostly use more than one device (i.e., many citizens use a laptop, a smartphone and a tablet).

One of the major challenges of the application of such a meta-app is the privacy issue, as will be discussed in Sect. 5.1. One expert stated the importance of storing user information on secure servers and of encrypting all information transmitted to the internet. Additionally, it must be ensured that communication with third-party providers is secure and trustworthy. An expert proposed to address the privacy issue as a possible unique selling point by clearly stating the purpose of the assembled (and shared) information.

Another important aspect, in the view of one expert, is that the app needs to run in the background of the phone's operating system and to periodically communicate with the server and the system.

One expert mentioned various features that could be added (e.g., information about current traffic, weather data and city news), and could be ordered and displayed by the preferences indicated by the users. A continuous synchronization of all these data is highly important for a proper functioning of the system. Additionally, it would be useful if the user could indicate the importance of the meetings so that the app can search for the fastest route and means of transportation to reach an appointment on time.

The experts provided constructive feedback, and these results are very helpful for the further development of the meta-app. Furthermore, we have to consider adapting this meta-app to other use cases, such as use by public authorities or a case without travel and to analyze other successful operating applications to observe how they function, what data they need, their pros and cons and what features are lacking yet are in demand.

5   Application of the Meta-App for Smart and Cognitive Cities

This section presents the application for smart and cognitive cities based on the paper prototype and the conducted evaluation. Overall, the presented meta-app can improve all implied perspectives of smart and cognitive cities. The combination of various aspects of city life can highlight the true meaning of smart and cognitive cities. The challenges of the application were also clearly demonstrated through the evaluation.

5.1 Perspectives, Challenges and Limitations

- *User Perspective*: The user perspective was tested in the evaluation and demonstrated the potential of the meta-app to enhance and optimize the daily lives of its users. The possibility to synchronize apps and the provided (data) privacy standards greatly augment the possible usage of the meta-app. The combination of the strengths and selling points of existing apps and the unique feature of assisted (i.e., ranging from manual to semi-automatic to automatic) decision making demonstrates the potential of the app.
- *Third-party Data Provider Perspective*: This perspective was addressed in the evaluation in a conceptual manner. The evaluation demonstrated various ranges of complexity to connect and synchronize the meta-app with current existing apps and future apps from third-party providers. The complexity is mainly technical but also business related and must address the question of willingness to share and provide data and interfaces to the meta-app.
- *City System Perspective*: The city system perspective was not the main focus of the prototype (e.g., the example of the evaluation was a dinner reservation). Nevertheless, the efficient usage of city infrastructure (e.g., roads and energy) as well as the city transportation facilities (e.g., trams, busses, and trains) was a main focus to save time. By saving time for an individual citizen using the city systems, all citizens benefit from this development. The optimized transportation of citizens through the provided transportation systems improves daily life in the city, reduces waiting times (e.g., in traffic jams, and from missed connections), reduces pollution because the transportation flow is stable.
- *New Perspectives*: After the evaluation further stakeholder groups and their requirements were discovered. The additional stakeholder groups are the following: Non-governmental organizations, non-profit organizations as well as crowdfunding. These perspectives have to be specified and analyzed in the next phase of development of the meta-app.
- *Challenges of the Next Phase of Design*: The application of the meta-app implies several challenges. An overview of upcoming challenges for a next design phase and future implementation is presented in Table 2. These challenges lead to requirements that will have to be defined in the next phase of the development process.
- *Information Representation*: The information flow in the meta-app, including all perspectives will be done through FCMs (see Sect. 2.2). In the current stage of development the information representation is not easily mediated to the user, as the underlying technical architecture of the meta-app is not clearly defined yet. Nevertheless, using FCMs to represent the proposed functionalities and the

Table 2 Upcoming challenges for the app

Challenge	Focus	Specifications
Data synchronization	Technical/participation	Willingness of other app/system providers to allow data synchronization
Data sharing	Technical/participation	Willingness of other app/system providers to allow data sharing
Privacy	Technical/participation	Willingness of the users to share their information
Security	Technical/participation	Encryption of the data and information
Data storage	Technical	How to store the data to make it available and usable
Data storage	Form	In which format/in which type of database will the data be stored
Functionalities	Technical/functional	Which functionalities to implement in the app
Decision making	Technical/functional	Willingness of the user to rely on the app for the decision making
Scale	Participation	Sufficient number of test users are needed to generate representative and reliable results
Synchronization with other sites	Technical/functional	What functional prerequisites are needed to make this possible?

addressed challenges for all stakeholders of the meta-app is key because FCMs are favored for this use [38].

- *Limitations*: As this chapter is focused on the evaluation of the paper prototype details, the explanation of the applied functions was not specifically extended by providing more thoroughly explanations. More information about the applied functions could be found in previous research papers [15–17]. Additionally, the evaluation of the paper prototype was performed but included several limitations. First, the number of involved experts and conducted interviews was small because this was the first user test outside of the researcher team [32]. Second, the applied paper prototype was established in a walk through, thus limiting the functional interaction from the user.

6   Conclusions

As today's cities continue to grow, methods of providing a good living environment must be developed. This goal can be attained by creating cognitive cities as an enhancement (i.e., through cognitive computing and connectivism) of smart cities. This chapter presents a tool that is based on fuzzy logic and cognitive computing and attempts to facilitate commuting and travel and thus optimizing the time schedules of citizens. This guarantees a good work-life balance and improves communication among stakeholders (e.g., citizens) because it automatically suggests new routes or means of transportation or arranges places for meetings.

This chapter presented an evaluation of a first prototype (i.e., paper prototype) of a meta-app for next generation communication in a cognitive city. An evaluation with five experts in three countries was conducted to test this paper prototype. The results of the evaluation show that the experts perceive the idea and the functionalities of the meta-app to be relevant and that further development should be pursued.

A next step in the development of the meta-app will be to implement these findings and thus improve the prototype. A future user test of the meta-app prototype will be conducted on a wider scale to be able to apply these findings in a software prototype including information representation through FCMs. A first version of the software prototype is under construction applying kansei engineering [31]. Meanwhile, the focus will also be to work on the technical challenges of the meta-app (e.g., security and privacy, data availability, and data storage) and on optimized stakeholder management. The software prototype represents a milestone for the meta-app as it will serve as proof of concept. Once this proof of concept is reached other areas of possible service applications (e.g., healthcare, public life, smart participation [44]) of the meta-app can be targeted.

Acknowledgments We would like to thank all of the experts who were willing to participate in the evaluation. Their feedback, based on their working experiences and knowledge, was helpful and without their participation, the improvement of the paper prototyping would not have been possible. We would also like to thank Sara D'Onofrio and Noémie Zurlinden, without whom the realization of this chapter would not have been possible. Additionally, we would like to thank the participants of the smart city project seminar (in collaboration with Accenture at the University of Bern) for their valuable insights.

References

1. Bailey, B.P., Biehl, J.T., Cook, D.J., Metcalf, H.E.: Adapting paper prototyping for designing user interfaces for multiple display environments. Pers. Ubiquit. Comput. **12**(3), 269–277 (2008)
2. Becker, G.S.: A theory of the allocation of time. Econ. J. 493–517 (1965)
3. Bolchini, D., Pulido, D., Faiola, A.: FEATURE Paper in screen prototyping: an agile technique to anticipate the mobile experience. Interactions **16**(4), 29–33 (2009)
4. Caragliu, A., Del Bo, C., Nijkamp, P.: Smart cities in Europe. 3rd Central European conference in regional science—CERS, (2009)
5. Clark, T., Osterwalder, A., Pigneur, Y.: Business Model You: A one-page method for reinventing your career. Wiley, Hoboken (2012)
6. Dearden, A.M., Naghsh, A., Ozcan, M.B.: Support for participation in electronic paper prototyping. In: Proceedings of the participatory design conference, pp 105-108. Malmo, Sweden, 23–25 June 2002. Palo Alto, CA, CPSR (2002)
7. Engelbart, D.C.: Augmenting human intellect: A conceptual framework Summary Report AFOSR-3223 under Contract AF 49 (638)-1024, SRI Project 3578 for Air Force Office of Scientific Research. http://www.dougengelbart.org/pubs/augment-3906.html (1962). Accessed 9 Dec 2014

8. Garrett, J.: The elements of user experience: User-centered design for the web and beyond. New Riders, Pearson Education, Berkeley (2010)
9. Goldstein, J.: Emergence as a construct: History and issues. Emergence **1**, 49–72 (1999)
10. Groumpos, P.: Fuzzy cognitive maps: Basic theories and their application to complex systems. In: Glyka, M. (ed.) Fuzzy cognitive maps: Advances in theory, methodologies, tools and applications, pp. 1–22. Springer, Berlin (2010)
11. Haubensak, O.: Smart cities and internet of things. In: Business aspects of the internet of things, seminar of advanced topics, pp. 33–39. ETH Zurich (2011)
12. Hundhausen, C., Fairbrother, D., Petre, M.: The "prototype walkthrough": a studio-based learning activity for the next generation of HCI education. In: Next generation of HCI and education: CHI 2010 workshop on UI technologies and educational pedagogy. Atlanta, GA (2010)
13. Hurwitz, J.S., Kaufman, M., Bowles, A.: Cognitive computing and big data analytics. Wiley, Hoboken (2015)
14. Johannesson, P., Perjons, E.: A design science primer. In: Create Space Publisher (2012)
15. Kaltenrieder, P., D'Onofrio, S., Portmann, E.: Enhancing multidirectional communication for cognitive cities. In eDemocracy & eGovernment (ICEDEG), 2015 second international conference on, pp. 38–43. IEEE (2015a)
16. Kaltenrieder, P., Portmann, E., Myrach, T.: Knowledge representation in cognitive cities, accepted for a special session of the IEEE international conference on fuzzy systems (FUZZ-IEEE), (2015b)
17. Kaltenrieder, P., Portmann, E., D'Onofrio, S., Finger, M.: Applying the fuzzy analytical hierarchy process in cognitive cities. In: Proceedings of the 8th international conference on theory and practice of electronic governance, pp. 259–262. ACM (2014)
18. Kaltenrieder, P., Portmann, E., Binggeli, N., Myrach, T.: A conceptual model to combine creativity techniques with fuzzy cognitive maps for enhanced knowledge management. Integrated systems: Innovations and applications, pp. 131–146. Springer, Berlin (2015c)
19. Kaufmann, M.A., Portmann, E., Fathi, M.: A concept of semantics extraction from web data by induction of fuzzy ontologies. In: International workshop on uncertainty reasoning for the semantic web (2012)
20. Kelly III, J.E., Hamm, S.: Smart machines, IBM's Watson and the era of cognitive computing. Columbia University Press, New York (2013)
21. Kosko, B.: Fuzzy cognitive maps. Int. J. Man Mach. Stud. **24**(1), 65–75 (1986)
22. Kosko, B.: Fuzzy engineering. Prentice Hall, New Jersey (1997)
23. Lewis, S.: The integration of paid work and the rest of life. Is post-industrial work the new leisure? Leisure Stud. **22**(4), 343–345 (2003)
24. Liu, L., Khooshabeh, P.: Paper or interactive? A study of prototyping techniques for ubiquitous computing environments. In: CHI'03 extended abstracts on human factors in computing systems, pp. 1030–1031. ACM (2003)
25. Medero, S., Cornell, K.: Paper prototyping. A list apart. Layout & Grids, Information Architecture, Interaction Design 231 (2007)
26. Meier, A., Stormer, H.: eBusiness & e-Commerce: managing the digital value chain. Springer, Berlin (2009)
27. Mendel, J.M., Zadeh, L.A., Trillas, E., Yager, R., Lawry, J., Hagas, H., Guadarrama, S.: What computing with words means to me. IEEE Comput. Intell. Mag. (2010)
28. Meuser, M., Nagel, U.: Das Experteninterview—konzeptionelle Grundlagen und methodische Anlage. Methoden der vergleichenden Politik- und Sozialwissenschaft, pp. 465–479. Neue Entwicklungen und Anwendungen. VS Verlag für Sozialwissenschaften, Wiesbaden (2009)
29. Modha, D.S., Ananthanarayanan, R., Esser, S.K., Ndirango, A., Sherbondy, A.J., Singh, R.: Cognitive computing. Commun. ACM **54**(8), 62–71 (2011)
30. Mostashari, A., Arnold, F., Mansouri, M., Finger, M.: Cognitive cities and intelligent urban governance. Netw. Ind. Q. **13**(3), 4–7 (2011)
31. Nagamachi, M.: Kansei engineering: a new ergonomic consumer-oriented technology for product development. Int. J. Ind. Ergon. **15**(1), 3–11 (1995)

32. Nielsen, J.: Usability engineering at a discount. In: Salvendy, G., Smith, M.J. (eds.) Designing and using human–computer interfaces and knowledge based systems, pp. 394–401. Elsevier Science Publishers, Amsterdam (1989)
33. Osterwalder, A., Pigneur, Y.: Business model generation: A handbook for visionaries, game changers, and challengers. Wiley, Hoboken (2010)
34. Papageorgiou, E.I., Salmeron, J.L.: A review of fuzzy cognitive maps research during the last decade. IEEE Trans. Fuzzy Syst. 66–79 (2013)
35. Pérez-Teruel, K., Leyva-Vázquez, M., Estrada-Sentí, V.: Mental models consensus process using fuzzy cognitive maps and computing with words 19(1), 7–22 (2015)
36. Portmann, E., Kaltenrieder, P.: The Web KnowARR Framework: Orchestrating computational intelligence with graph databases. Information granularity, big data, and computational intelligence, pp. 325–346. Springer, Berlin (2014)
37. Portmann, E., Kaufmann, M.A., Graf, C.: A distributed, semiotic-inductive, and human-oriented approach to web-scale knowledge retrieval. In: Proceedings of the 21st ACM international conference on information and knowledge management. Maui, Hawaii, USA (2012)
38. Portmann, E., Kaltenrieder, P., Pedrycz, W.: Knowledge representation through graphs. In: Proceedings of the 2015 international conference on soft computing and software engineering (SCSE'15), Procedia Computer Science, vol. 62, pp. 245–248. (2015)
39. Ruh, H.: Anders, aber besser: Die Arbeit neu erfinden—für eine solidarische und überlebensfähige Welt 3. Auflage. logo, Waldgut (2002)
40. Sefelin, R., Tscheligi, M., Giller, V.: Paper prototyping—what is it good for? A comparison of paper- and computer-based low-fidelity prototyping. In: CHI'03 extended abstracts on human factors in computing systems, pp. 778–779. ACM (2003)
41. Siemens, G.: Connectivism: A learning theory for the digital age. Int. J. Instr. Technol. Distance Learn. 2(1), 3–10 (2005)
42. Siemens, G.: Knowing knowledge. Lulu.com (2006)
43. Signer, B., Norrie, M.C.: PaperPoint: a paper-based presentation and interactive paper prototyping tool. In: Proceedings of the 1st international conference on tangible and embedded interaction, pp. 57–64. ACM (2007)
44. Tamayo, L.F.T.: SmartParticipation: a fuzzy-based recommender system for political community-building. Springer (2014)
45. Thomas, K.: Work and leisure. Past Present 50–66 (1964)
46. Tolman, E.C.: Cognitive maps in rats and men. Psychol. Rev. 55(4), 15–64 (1948)
47. Townsend, A.M.: Smart cities: big data, civic hackers, and the quest for a new utopia. WW Norton & Company (2013)
48. Snyder, C.: Paper prototyping: The fast and easy way to design and refine user interfaces (interactive technologies). Morgan Kaufmann Publishers, Amsterdam (2003)
49. United Nations Department of Urban and Social Affairs: World urbanization prospects: The 2007 revision—executive summary. http://www.un.org/esa/population/publications/wup2007/2007WUP_ExecSum_web.pdf (2008). Accessed 8 June 2015
50. Wing, J.M.: Computational thinking. Commun. ACM 49(3), 33–35 (2006)
51. Zadeh, L.A.: Fuzzy sets. Inf. Control 8(3), 338–353 (1965)
52. Zadeh, L.A.: Fuzzy logic. Computer 21(4), 83–93 (1988)

Verbalization of Dependencies Between Concepts Built Through Fuzzy Cognitive Maps

Marcel Wehrle, Marc Osswald and Edy Portmann

Abstract The new computing paradigm known as cognitive computing attempts to imitate the human capabilities of learning, problem solving, and considering things in context. To do so, an application (a cognitive system) must learn from its environment (e.g., by interacting with various interfaces). These interfaces can run the gamut from sensors to humans to databases. Accessing data through such interfaces allows the system to conduct cognitive tasks that can support humans in decision-making or problem-solving processes. Cognitive systems can be integrated into various domains (e.g., medicine or insurance). For example, a cognitive system in cities can collect data, can learn from various data sources and can then attempt to connect these sources to provide real time optimizations of subsystems within the city (e.g., the transportation system). In this study, we provide a methodology for integrating a cognitive system that allows data to be verbalized, making the causalities and hypotheses generated from the cognitive system more understandable to humans. We abstract a city subsystem—passenger flow for a taxi company—by applying fuzzy cognitive maps (FCMs). FCMs can be used as a mathematical tool for modeling complex systems built by directed graphs with concepts (e.g., policies, events, and/or domains) as nodes and causalities as edges. As a verbalization technique we introduce the restriction-centered theory of reasoning (RCT). RCT addresses the imprecision inherent in language by introducing restrictions. Using this underlying combinatorial design, our approach can handle large data sets from complex systems and make the output understandable to humans.

M. Wehrle (✉)
University of Fribourg, Fribourg, Switzerland
e-mail: marcel.wehrle@unifr.ch

M. Osswald
Swiss International Air Lines Ltd., Basel, Switzerland
e-mail: marc.osswald@swiss.ch

E. Portmann
University of Bern, Bern, Switzerland
e-mail: edy.portann@iwi.unibe.ch

© Springer International Publishing Switzerland 2016
E. Portmann and M. Finger (eds.), *Towards Cognitive Cities*,
Studies in Systems, Decision and Control 63, DOI 10.1007/978-3-319-33798-2_7

Keywords Fuzzy cognitive map · Restriction centered theory · Verbalization · Natural language · Fuzziness · Graph database

1 Building Cognitive Cities with Cognitive Computing

Transportation systems today are faced with a variety of challenges stemming from population growth and higher customer expectations, which raise issues such as congestion, longer commutes, and the inadequacy of public transportation [16]. In 2050, the OECD expects the number of vehicles worldwide to double to 2.5 billion [11]. Therefore, several optimizations are in development to enhance transportation systems' capabilities, such as the integration of the Internet of Things (IoT), where every transportation object has a unique IP address. Cities today already use data from sensors placed in roads and on vehicles to anticipate traffic loads [2]. Detecting users' daily transportation behaviors and matching them with the available and corresponding public transportation options can help to reduce road congestion. At first glance, it might appear that these problems can be handled by augmenting operational efficiency. However, transportation systems in cognitive cities interact directly with customers to optimize various aspects of the system (e.g., by controlling the scale of commuting activities during peak hours). These systems try to combine all the accessible information and then extract the information required to optimize coordination and communication between the transportation objects. The information flow of transportation systems from objects to users is bi-directional; however, clear communication is a challenge because natural language is not easily interpretable for machines. Thus, there is a need for computing methods that introduce cognition into transportation systems. Cognitive computing, a term inspired by Watson from IBM [9], might be a solution to transform data into valuable information that is understandable by both humans and machines. Unfortunately, because human language is often vague and imprecise, fuzzy computing methods must be employed. A few such methods have already been introduced toward this goal under the umbrella term *computing with words*, which is nothing less than a methodology for computing with natural language, inspired by the human capability to perform various tasks without any need for measurement or computation [20]. Computing with words can build bridges between machines and humans and take advantage of the strengths of both to simplify the communication process.

1.1 From Smart to Cognitive Cities

Cities, especially large metropolises, will be forced by increasing problems caused by, among other things, steady demographic growth and the continuing population shift from rural areas to cities. According to the UN, 66 % of the world's population will live in urban areas by 2050 (compared to 54 % today), when the world's population will be 9.7 billion. This shift is leading to what are called megacities such as Tokyo, Delhi, and Shanghai [19]. The urban population has grown significantly in the last 50 years, and this process will continue over the next 50 years. The principles (e.g., Smart Living, Smart Mobility or Smart Energy) underlying smart cities can help solve the social, economic, and environmental problems caused by these trends. In one sense, these cities are places where information technology meets objects in the form of human capital, social capital, or information, with the goal of generating greater and more sustainable economic development and a better quality of life [5]. Meeting this goal involves collecting, analyzing, and applying data from various objects belonging to a city. In fact, there are already many (European) initiatives to make cities "smarter" [5]; however, the meaning of "smart" is not clearly defined. Is it smart when a system knows when you are at home so it can adjust the heat setting in your apartment? Depending on your perspective, it is. Suppose you add more objects to your personal environment? What if your environment were not simply driven by rules, but actively learned from you? What if your environment were connected with your friends' environments? For instance, your environment might know exactly where you and your friends are at the moment as well as all the possible timeframes when you might be able to meet them. These questions are indicative of the steps that systems must take to move from smart, to learning, to cognitive environments. Connecting these concepts leads to cognitive cities. In cognitive cities, no device (e.g., mobile phone) or building has center focus; instead, the system consists of humans surrounded by a network of things (the Internet of things)—including such things as buildings and phones. The objects in the network communicate with one another in context to deliver a better experience for humans (e.g., by augmenting reality to expose the content interesting to humans).

Figure 1 shows the steps involved in moving from smart to cognitive environments/cities as well as the added value at each step. Cognitive cities are a progression from smart cities. They are also described by other terms, including *intelligent cities*, *digital cities*, and *cyber cities* [17]; all of these terms signify that different characteristics of the cities but are complements of smart cities. Smart cities attempt to optimize and channel existing resources by interconnecting all the available

Fig. 1 From smart to cognitive cities

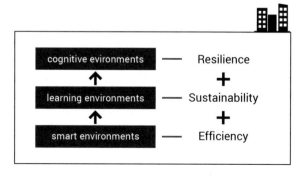

information related to the model. There are various manifestations at more detailed levels that try to improve systems already in place. In cognitive cities, a system not only optimizes but also learns from analyzing the flow of data. It can access data and provide solutions in real time to improve human decision-making capabilities [7].

1.2 Cognitive Systems

By applying the principles of cognitive computing, humans can extend their expertise to interactions with machines to provide enhanced support for their daily lives. Applying cognitive computing principles leads to cognitive systems that can derive meaning from natural language, speech, or text in documents to generate and evaluate hypotheses rapidly based on huge amounts of data [6]. By storing and connecting meaning, cognitive systems can create (new) knowledge and learn from human-generated content.

As its first step, a cognitive system needs a base foundation of information in a given domain, a so-called corpus. The input necessary to create the corpus can include data from sensors, images, speeches, or texts. Using various learning procedures, the corpus can add new information to provide better hypotheses. Hypotheses are built and scored using models that define how data are analyzed and which algorithms are used in the corresponding context by identifying patterns. Cognitive systems should learn from their experiences through interactions with different interfaces, including with human beings. Another important point is the communication between interfaces. A cognitive system is able not only to analyze natural language in context but also to communicate in natural language; hence, it simplifies communications with humans [4] and provides a more natural experience.

Cognitive computing is considered an extension or a logical development of semantic computing. In fact, both are trending in information science, and differentiating between them is not a simple task. Often semantics and cognitive computing are discussed in relation to smart computing. Cognitive computing requires semantic computing as a substrate to process data [14].

1.3 Applying Soft Computing in a Cognitive System

Based on the components that cognitive systems need, as discussed in Sect. 1.2, in this section we propose fuzzy cognitive maps as a possible model as well as the use of restriction centered theory as a tool to better translate the hypotheses generated. Broadly speaking, cognitive maps can be drawn between entities belonging to cities. These can be various things—buildings, streets, humans, mobile phones, among others—the only important thing is that they deliver useful data to the corpus of a cognitive system. These entities are connected to one another based on dependencies. These dependencies can vary in meaning or strength; for example, they can represent cities' functional dependencies on natural resources. This method can be applied to many domains. In this case, for example, a restriction is that city entities are considered connected nodes.

Fuzzy cognitive maps (FCMs) are a simple and powerful approach to linking nodes or concepts together. Such relationships can be weighted normally from −1 to 1 to provide more information and to introduce weighted dependencies. Using this approach, a vast number of models can be formed. Notably, FCMs are capable of learning from changing environments. When applied to the field of transportation, these relationships can be interconnections of cities for exchanging people or goods. Nodes may be the cities themselves, or more specifically, airports, bus or train stations, harbors, and so on. Experts usually model cognitive maps so that the relationships they contain can be easily and logically named. This is also possible for machines with the help of cognitive computing.

In Sect. 2, FCMs are introduced and their designated use is outlined. By applying learning algorithms such as Hebbian learning, FCMs can iteratively gain knowledge. The learning methods mentioned will be presented and discussed, especially regarding their applicability to big data. After this first step, the problem remains that the purely numerical output that FCMs output as hypotheses are difficult for humans to understand. The challenge here is to distinguish between essential and dispensable information and, in addition, to create statements that are correct in a human language. This task is clarified by defining restriction-centered theory (RCT), which can be used by FCMs for verbalization. RCT has a fuzzy approach and is basically used to turn natural language into a computable format. However, for purposes of this study, RCT must be applied in reverse—that is, to form sentences out of computable or computed contents. The possibility of doing so is outlined in this paper. Additionally, the potential of applying RCT to large amounts of data will be discussed.

RCT is primarily based on the work of Zadeh [21], whose paper is the starting point for transforming natural language into computable input. The inverse direction—the transformation of mathematical values into natural language in which the computable elements are contained in an FCM—is then proposed. As a proof of concept, an FCM will be developed with live data, and the numbers will be translated into natural language. After this, the truth and suitability of the resulting statements will have to withstand the challenge of a qualitative expert.

The findings will be tested in a use case: passenger data for a taxi company. Those data will be included in an FCM and hypotheses in natural language format will be produced based on the same data. For reasons of simplicity, this is a rudimentary application of the method. In a real cognitive system many more data sources would be included in the corpus than just passenger data. Integrating more information in context is left for a future work.

2 Soft Computing Methods for Cognitive Computing

This section first provides an overview of how FCMs and the RCT can be integrated into a cognitive system. Then, the basics of FCMs and the RCT and how it is possible to combine them are covered. Finally, the interplay of the different elements in a cognitive system is discussed.

2.1 *Introduction*

Soft computing methods, such as computing with words, are widely used, especially in domains in which systems have to cope with uncertainty and imprecision. Soft computing's underlying principles lie in fuzzy logic theory. Fuzzy logic is a renunciation of classical quantity-based operations; through membership functions, elements can be assigned to multiple sets. This allows computers to handle events that are not one hundred percent certain. Soft computing methods (e.g., artificial neural networks or evolutionary computing) are used in domains such as robotics, artificial intelligence, and machine learning. In this section, the soft computing methods of FCM and RCT are integrated into a cognitive system. Both cognitive computing and soft computing use the human brain as a paradigm to execute complex tasks that support and amplify human capabilities in daily life. Because humans often have to decide based on vague fundamentals, it is reasonable to introduce methods that can cope with uncertainty. Humans learn based on their already available knowledge through the processes of observation, rating, explaining, and deciding. The proposed method for a cognitive system makes use of similar processes. FCMs correlate to the model of a cognitive system and the RCT provides services to translate hypotheses into natural language. Figure 2 shows the basic architecture of our proposed system.

Cognitive systems can learn from new data input. For example, in a cognitive map of the taxi transportation system of a city, the corpus might contain data about

Fig. 2 Architecture of the
cognitive system

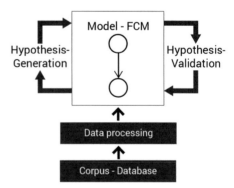

the number of taxis (e.g., traffic information, hotspots, streets, intersections, passengers, and passenger priorities). Most of these data consist of numbers provided to the RCT, which uses them as a tool to verbalize the generated hypotheses and leads to better communications between the cognitive system and the humans who are connected with it. Going a step further, the system itself can also deverbalize human language to generate input—the numbers that the system understands.

2.2 Fuzzy Cognitive Maps

The widely known "mind map" is at the root of an FCM [1]. A mind map begins with a central term or concept. Other terms associated with the initial term are written around the first one and linked to one another with graphs. These associated terms may be used as central nodes as well. In the end, a tree structure is formed [13]. It is obvious that most concepts cannot be described using tree structures exclusively. As a simple example, consider the road (or railroad) network of a country. If the conception were created according to a mind map, there would be one capital city (probably the most populous) as the central point. The next biggest cities would be linked to the capital city with roads, and smaller towns and villages would be linked to these cities.

The road networks of nations such as France have a centralistic structure that conforms to this method of conceptualization. Many roads lead to Paris, the capital. However, there is clearly a need for links between smaller towns as well, so that someone who wants to get from Lyon to Marseille is not forced to make a detour through Paris.

These direct interconnections expand the tree structure of the mind map with an important element. Links between any elements of the map can be represented this way. Consequently, the mind map is no longer a simple tree structure; cycles can occur. This concept is much closer to reality and is a huge step toward cognitive mapping. Cognitive maps can interconnect not only roads but also any sort of precise or abstract concepts. The concepts are represented by nodes and the interconnections by directed graphs [15]. Additionally, the edges are used to represent information about how one concept influences another. This information is reduced to a plus or minus value, meaning that a concept exerts a positive or a negative influence on its counterpart. The last step toward building an FCM is the introduction of edge weights, which have values in the interval between −1 and 1. Edge weights make it possible to present a statement about how much one concept influences another.

According to Stach et al. [18], an FCM can be developed either manually or computationally. For example, the FCM in Fig. 3 represents cognitively how the (future) traffic of a taxi enterprise can be influenced. As this example shows, information can be illustrated in a trivalent way. In other words, the absence of a graph bears the same informational value as graphs with their plusses or minuses.

Technically speaking, this map is a special form of an FCM [15]. Usually, the weights of an FCM's graphs are not measured in discrete numbers (e.g. 1, 0, or −1) but

Fig. 3 Fictive cognitive map
of influences taxi route A to B

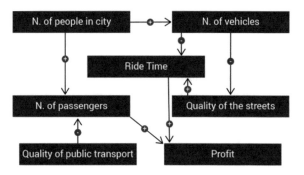

can hold any value in the interval $[-1, 1]$. This increase of granularity supports an illustration of the strength of influences between concepts. In the words of Kosko [10], "Fuzziness allows hazy degrees of causality between hazy causal objects (concepts)."

2.2.1 Formal Definition of an FCM

The following formal definition was developed by Stach et al. [18]. It allows for the possibility of dynamically calculating an FCM using iterative steps. Let $\mathbb{R}$ be the set of real numbers, while $\mathbb{N}$ denotes the set of natural numbers. $K = [-1, 1]$ and $L = [0, 1]$. A fuzzy cognitive map F is a quadruple (N, E, C, f), where:

1. $N = \{N_1, N_2, \ldots, N_n\}$ is the set of n concepts forming the nodes of a graph.
2. $E : (N_i, N_j) \rightarrow e_{ij}$ is a function of $N \times N$ to K associating e_{ij} to a pair of concepts (N_i, N_j), where e_{ij} denotes the weight of a directed edge from N_i to N_j, and where $i \neq j$ and e_{ij} is equal to zero if $i = j$. Thus $E(N \times N) = (e_{ij}) \in K^{n \times n}$ is a connection matrix.
3. $C : N_i \rightarrow C_i$ is a function that at each concept N_i associates the sequence of its activation degrees. For example, for $t \in \mathbb{N}$, $C_i(t) \in L$, given its activation degree at the moment t, $C(0) \in L^n$ indicates the initial vector and specifies the initial values of all concept nodes, and $C(t) \in L^n$ is a state vector at a certain iteration of t.
4. $f : \mathbb{R} \rightarrow L$ is a transformation function that includes the recurring relationship for $t \geq 0$ between $C(t+1)$ and $C(t)$.

An FCM consists of both a static part and a dynamic part. The nodes and edges with their weights, as well as the transformation function, are static, whereas the state vectors change with every iteration step. What is missing is an initial vector, the transformation function and, finally, an instruction as to what must be completed to iterate the FCM. This last element is called the functional model.

A simulation run on an FCM begins with a scenario that is expressed in an initial vector. This vector is included in the functional model. The result of the calculation step is a new vector, called the state vector. The relationship between the initial vector and the iteration result is a cause-and-effect statement. When iterating the FCM several times, it is possible for a pattern to occur. The sequence of state vectors then repeats. When the state vector no longer changes, the FCM is in a steady state. On the other hand, it is also possible for a new state vector to be produced with every iteration. This type of FCM is called chaotic.

The functional model described by Stach et al. [18] is as follows:

$$\forall i \in \{1,\ldots,n\}, C_i(t+1) = f\left(\sum_{\substack{i=1 \\ i\neq j}}^{n} e_{ji}C_j(t)\right) \tag{1}$$

Functional model of an FCM

In this way, the initial vector is multiplied with the vector of corresponding edge weights. Because these weights are supposed to be fuzzy, a transformation function is used to keep the values within a certain range defined at the beginning. There are many different ways to normalize the values; the most common, according to Stach et al. [18], are the following:

$$\text{bivalent:}\quad f(x) = \begin{cases} 0, & \textit{if } x \leq 0, \\ 1, & x > 0 \end{cases} \tag{2}$$

Bivalent transformation function

$$\text{trivalent:}\quad f(x) = \begin{cases} -1, & \textit{if } x \leq -0.5, \\ 0, & \textit{if } -0.5 < x < 0.5, \\ 1, & \textit{if } x \geq 0.5 \end{cases} \tag{3}$$

Trivalent transformation function

$$\text{logistic:}\quad f(x) = \frac{1}{1+e^{-cx}}, c \in \mathbb{R} \tag{4}$$

Logistic transformation function

For the trivalent and the logistic transformation functions, the formal definition above has to be rewritten because the range of L is the interval between 0 and 1. The scope of findings that can be drawn out of such a simulation is very narrow. Therefore, another improvement to the FCM will be introduced in the following section.

Several key indicators can be used to interpret an FCM at first glance. According to Özesmi and Özesmi [12], a node can be classified into three types: transmitter, receiver, and ordinary. The determination is based on two key figures, outdegree and indegree. The outdegree od_i of a node N_i is the sum of its outgoing absolute edge weights, while the indegree id_i of node N_i sums the absolute weights of all incoming edges.

$$od_i = \sum_{j=1}^{n} |e_{ij}| \tag{5}$$

Outdegree of a node

$$id_i = \sum_{j=1}^{n} |e_{ji}| \tag{6}$$

Indegree of a node

If a node has a positive outdegree and a zero indegree, it is a transmitter node. If it has a zero outdegree and a positive indegree, it is a receiver node. If both degrees are positive, it is an ordinary node. Nodes with zero indegree and zero outdegree obviously do not play any role in an FCM. The relevance of a node within the FCM can be expressed with the centrality td_i, which is simply the sum of its outdegree and indegree. The higher the centrality number of a node, the greater its influence on the FCM. Finally, the connectivity of the entire FCM can be measured using the density d, which is the ratio of existing connections to all possible connections.

$$td_i = od_i + id_i \tag{7}$$

Centrality of a node

$$d = \frac{E}{N(N-1)} \tag{8}$$

Density of an FCM

2.2.2 Learning Algorithms

It is often problematic to argue about the edge weights of an FCM. Typically, they are determined by human interaction. Depending on the field of research, this process is heavily subjective. One approach to solving the problem is through learning algorithms. As was mentioned in Sect. 2.2.1, an FCM consists of a static and a dynamic part. This section discusses the change in edge weights from the static to the dynamic part, so that the edge weights initially defined by experts can be computed. This increases both the complexity as well as the range of findings that can be extracted from an FCM.

Differential Hebbian Learning
There are many different ways to learn by modifying edge weights. A first attempt was made by Dickerson and Kosko [3], who proposed a differential Hebbian learning law that could be applied to FCMs:

$$\dot{e}_{ij} = -e_{ij} + \dot{C}_i \dot{C}_j \tag{9}$$

Differential Hebbian learning law

In this equation, $\dot{e}_{ij}$ stands for the change of weight between nodes N_i and N_j, e_{ij} is the current weight, and $\dot{C}_i$ and $\dot{C}_j$ are changes of the activation degree of nodes N_i and N_j. Expressed in a discrete way, the updated equation is defined as:

$$e_{ij}(t+1) = \begin{cases} e_{ij}(t) + c_t \left[\Delta C_i \Delta C_j - e_{ij}(t) \right] & \text{if } \Delta C_i \neq 0, \\ e_{ij}(t) & \text{if } \Delta C_i = 0 \end{cases} \tag{10}$$

Discrete update equation for DHL law

Here, $e_{ij}(t)$ denotes the weight between nodes N_i and N_j at the moment t, ΔC_i and ΔC_j are the changes of activation degrees of nodes N_i and N_j, and c_t is a decreasing learning coefficient. It is notable that the edge weights are updated only if the state value of the outgoing node N_i changes. As a result of the decreasing learning coefficient, the learning effect is lowered with every iteration. The concepts move in the same direction if ΔC_i and ΔC_j are both either positive or negative. In these cases, the edge weights move toward the minimum (-1) or maximum (1) weight. When concepts move in different directions, the algebraic signs of ΔC_i and ΔC_j are not equal. In this case, the edge weights move toward zero, which is in the middle of the value range. The next section introduces RCT is introduced, which is used to verbalize the connections produced by FCM to make them more understandable for humans.

2.3 Restriction-Centered Theory

The restriction-centered theory was developed by Zadeh [21]. The main difference between common logical systems and the system described in the RCT lies in the variety of valence. Traditional logical systems discriminate between true and false (between 0 and 1). In the RCT, however, a much higher variety exists, as the entire range between 0 and 1 is available. In other words, RCT encompasses everything from true to not true, including all the gradations such as very true, really true, quite true, a bit true, hardly true, etc. A traditional logical system answers binary questions such as "Is X true?", whereas the RCT provides a clue as to how true X is. Unless otherwise noted, the content of this chapter is based on Zadeh [21].

2.3.1 General Idea of the RCT

In general, words and natural language are very important to the RCT. Using a word, a human can express fuzziness in a much simpler way than by using probabilistic models. The challenge is to transform these words into a computable form. In brief, this is what the RCT seeks to do. If the answer to the question "How true is

X?" is expressed as a value, the range of that value is between 0 and 1 (the range can be different but is always defined). The same concept is valid for a set with predefined values or a probability distribution. Each of these concepts has a homogeneous range. A word can be used in two different ranges—that is, in two different contexts. Worse, the same word can have different meanings according to the context in which it is used.

The central concept of the RCT is restriction, which can be implemented in a more general way than an interval, a set, or a probability distribution. The range of a restriction is a melting pot with all the concepts mentioned, and it can take any value. To understand the size of this range, here is an example: If a human is asked a question such as "How tall is Steve?", some possible answers might be "Steve is 1.85 m tall," "Steve is 6 feet and 1 inch tall," "Steve is quite tall," "Steve is taller than Angelica," "Steve is not as tall as Michael," "Steve is between 5 and 10 cm taller than I am," and so on. The first two statements do not leave any room for ambiguity, but the last four of these restrictions are theses, which Zadeh [21, p. 2] called propositions. They are drawn from natural language and usually contain a fuzzy component such as a:

- predicate (small, strong, slow)
- quantifier (lots, little, more)
- and/or a probability (likely, unlikely)

A distinction is made between zero-order and first-order fuzzy propositions: The former contain only a fuzzy predicate, whereas the latter contain any one or more of the components mentioned. A computer asking how tall Steve is can interpret only the first two answers, unless it is able to turn the proposition into a computable form. With the RCT, this turns out to be feasible.

To represent a proposition made in natural language in a mathematical form, the meaning postulate (MP) is used:

$$p \to X \, isr \, R \tag{11}$$

Representation of a restriction as a meaning postulate (MP)

In this equation, p is the proposition that contains a variable to be restricted, X, a restricting relation, R, and the type of restricting relation, r. The better the mathematical definitions of X, R, and r are, the better a restriction can be computed. Computations with restrictions look like this example with fictive numbers: "One out of about 100,000 people is affected by the genetic disorder called maple syrup urine disease. In Switzerland, some 20 people are affected by this disease. What is the population of Switzerland?" These two restrictions imply that Switzerland has only two million inhabitants. Humans are used to dealing with this type of fuzzy reasoning. RCT offers an approach for solving such problems computationally.

A second important element of the RCT is the canonical form (CF). Basically, $CF(p)$ is the right-hand side of (11), which assigns the correct restriction type to the proposition. The last element to be mentioned is the truth postulate (TP), which measures the truth degree of an MP and is closely linked to the preciseness of the proposition. The truth degree can be expressed either numerically, called a first-order truth value, or in natural language, which is a second-order truth value.

2.3.2 Restrictions

A restriction is called singular if the type of the restricting relation, R, is a singleton (e.g., $X = 5$). It is called nonsingular if R is not a singleton. The restricted variable may be X or a function of X. In the first case, the restriction is called direct; otherwise it is indirect. A restriction can be differentiated into types. The three most important types are the possibilistic restriction, the probabilistic restriction, and the Z-restriction, which is a combination of the first two types.

In the possibilistic restriction, R is a fuzzy set, A, to which X belongs. The affiliation of X to A is evaluated with the membership function, μ_A, of a possibility distribution. The same is valid for fuzzy relations where two variables are compared. For example, A is a fuzzy set stating whether someone is young. The membership function is defined in such a way that everyone who is younger than 25 belongs completely to this set. Anyone older than 50 does not belong to the set at all. In between these two limits, a linearly declining curve gives an idea of how young someone is. When facing the proposition "Person A is young," the possibility of $X = Age(Person\,A) = 30$ can be measured using the input given. Moreover, the comparative proposal, "Person A is younger than Person B," can also be measured.

With probabilistic restriction, a statement about the certainty of a proposition can be made. This is evaluated using a density function of X. Applied to statements with natural language, this means that the certainty of "usually" is higher than that of "sometimes." It is rare for probabilistic restrictions to occur exclusively in natural language; they are more often found in combination with possibilistic propositions. This combination is called a Z-restriction; a Z-restriction incorporates natural language propositions such as "Maybe person A is young," where "young" is possibilistic and "maybe" is probabilistic.

The notation of the CF differs depending on the type of restriction. A possibilistic restriction is denoted as X is A, where A is the fuzzy set or relation. A probabilistic restriction is shown as X is p, where the first p is the probability density function (not to be mistaken with the second p for proposition in the MP). A Z-restriction is indicated as X iz Z, where Z is the combination of possibilistic and probabilistic restrictions defined as $Z : Prob(X \text{ is } A) \text{ is } B$.

In most cases, the restriction is formulated in natural language. The process of transforming the linguistic input into a computable form is called precisiation. To achieve this, a so-called explanatory database (ED) is used, which is a collection of relations containing the data source to precisiate the variables X and R or to compute the truth value of p. When the variables are precisiated, they are denoted by X^*, R^*, and p^*. So, $X^* = f(ED)$ and $R^* = g(ED)$. The numerical truth value of p, nt_p, can be expressed as $nt_p = tr(ED)$, where tr is referred to as the truth function.

2.3.3   RCT on FCMs

It is now shown how natural language can be turned into a computable input. If there exists an FCM with nodes, edges, and their respective weights, the computable input is available but cannot be interpreted by an inexperienced person. Therefore, the problem is to convert the numerical input of an FCM into natural language with the aid of the RCT (Fig. 4).

First of all, it must be made clear how the RCT can make use of an FCM: The FCM is usually built based on experiences, whether those are the thoughts of experts and observations or some other large data source. Essentially, the FCM serves as the ED for the RCT because it provides the data that are used to precisiate variables. To build a sentence grammatically, a subject and a predicate are the minimum requirements. As a subject, a certain node of the FCM can be utilized, regardless of context, but the predicate must be defined by the user, depending on the statement that he or she wants to make. But even after selecting a subject and predicate, the budding sentence is not yet very informative. Therefore, two further elements must be added to create a standard sentence: an object, where another node can be placed, and a descriptive word such as an adjective or adverb. It is conceivable that the edge weights can be used for this purpose because they provide values concerning the strength of a relationship between the subject and the object. Depending on the weight, a certain word can be chosen to describe the content.

At this juncture, the RCT comes into play. The choice of words is similar to a possibilistic restriction. Each word has a different membership function. The totality of word options should cover the full range of edge weights. The membership degree of each eligible word then must be evaluated and the most appropriate among them chosen. If there are two or more words with the same membership degree, the system assumes that those words can be used similarly. Statements of different natures can also be made based on the key figures presented in Sect. 2.2. Additionally, the state of a node at a certain moment bears instructive information. Overall, the nature of the statements that can be made strongly depends on the content of the FCM. This context dependency is what makes it difficult to define, for example, clear sets of words that can be used for any purpose. A cognitive system's ability to learn allows these sets of words to be regarded in context. The following section defines a use case to provide a clearer notion of what such a system might look like.

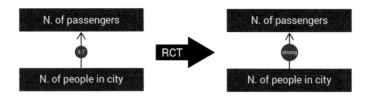

Fig. 4 Transformation of numerical to verbalized output using RCT

2.4   The Interplay of the Elements

An FCM cannot be used on every dataset; there is a strong need to create cognitive connections between the elements that can be translated into numbers. This connection can be influenced by a large number of components, depending on the use case. For example, for a traffic situation from point A to point B, these cognitive connections could include the number of vehicles, the weather, the state of the streets, etc. To summarize, an FCM usually allows the transformation of complex data into a cognitive map to support various decision-making processes. In this example, the decision might be to choose the best route to travel from point A to point B. In contrast, the RCT is a method that supports translating these numerical connections so that they are more understandable to humans. The translation makes the decision-making process faster, and no additional knowledge is needed to interpret the connections. Of course, the RCT's translations must be observed in context; not every natural language statement is understood equally by all humans. For some people, for example, a temperature of 15 °C is quite agreeable, whereas for others it is very cold. Identifying this context is precisely one of the strengths of (future) cognitive systems.

By combining FCM and the RTC into a model, properties evolve that are suitable for cognitive systems. The model can handle functional and logical dependencies and translate them so humans can understand the system's hypotheses. In addition, this makes it possible for humans to communicate with the model in their own languages, which fits perfectly with the ideas behind cognitive cities. Such a model can handle large amounts of data; in fact, it becomes more useful as the amount of data grows. Now, the main remaining limitation is mapping natural language because that step is not performed automatically and still requires experts to verify the output of the model. These are also problems in cognitive systems.

3   Use Case: Taxi Passenger Flows

3.1   Introduction

The following use case will investigate the flow of passengers for a taxi company. Based on passenger flows, an FCM will be constructed, and then a simulation will be run. Finally, the result will be interpreted by a tool called an interpretation engine, which converts the output into natural language hypotheses.

3.2   Data Structure

Data analysis plays an obvious role in the taxi business. There are many important logistical questions to answer, such as: Where should taxis go to pick up as many customers as possible? Where are those customers going? On one hand, the answers to these questions can be used by the taxi company in deciding where to send its drivers, and on the other hand, can be used by the government to estimate the required size of taxi waiting areas for city planning purposes.

Using existing data, it is possible to provide a general impression about passengers' journeys from boarding to disembarkation. For the sake of simplicity, the system does not consider every address; instead, it considers the most notable points of interest in the city. If a customer is heading to a specific address, the system considers the nearest point of interest. This leads to a passenger drift towards points of interests that are either located in densely populated areas or are far from other points of interest.

3.3   Creation of the FCM

Before transforming any data, the structure and meta-information of the FCM must be defined. This will be performed in the first subsection. Afterward, the transformation steps are detailed.

An FCM consists of three components: nodes, graphs, and properties. When designing a new FCM, the specifications of these components must be determined before importing the data. The first problem lies in specifying these nodes. In this scenario, the nodes can be depicted as points of interest or as routes.

The nodes represent points of interest; a passenger traveling from the zoo to the railway station would be depicted as a direct link, as shown in Fig. 5. The most important places of the city in this case study are the airport, casino, cemetery, convention center, hospital, indoor swimming pool, night club, outdoor swimming pool, port, railway station, retirement home, stadium, theater, town hall, shopping center, university, and the zoo. In addition, several important business locations for travelers—hotels, museums, and tourist destinations—are considered, as well as the four most important suburbs in the city. The less important considerations are summarized under other companies, other hotels, etc.

Fig. 5 Specification of FCM nodes as points of interest

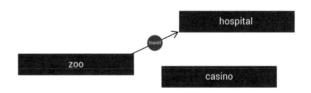

By making decisions along these lines, 42 points of interest are available in the present case study. The specification of the graphs is straightforward: A graph must be created between A and B if a passenger travels from A to B. The graph points in the direction of travel. The properties of the new FCM's elements are as yet unspecified. Nodes hold information such as their name—that is, the name of the point of interest—description (including its location coordinates), and finally, the total number of passengers boarding from that location. Graphs hold the information for the route a passenger will travel. Additionally, the absolute and relative number of passengers on a graph is provided.

3.4 Query Engine

To interpret the FCM, a query engine must be defined that generates sentences in natural language. These statements deliver deeper insight into the FCM and allow a simpler interpretation of the data. For the present use case, the following issue is considered: the frequency of passengers between two specific points in a directed sense.

The frequency of passengers from one point, A, to another, B, in relation to the total number of passengers boarding at any point is very low. There are 1554 possible different connections between two points; the median is 0.04 %, and the 90th percentile is 0.1 %, which indicates that most of the connections have a very low share and explains why it is important to have an accurate distinction for low shares. In contrast, the shares above 10 % can be described with only a few different words. Therefore, connections between the 10 and 100 % range will be described with five different adjectives. The remaining range between 0 and 10 % will be covered with nine different adjectives.

All these adjectives are intended to express the frequency of a connection and are therefore ordered according to their frequency: never, seldom, rarely, occasionally, infrequently, sometimes, frequently, often, regularly, normally, usually, generally, hardly ever, and always. Each of these words has a membership function that defines the word's degree of truth in relation to the share of connecting passengers. All the membership functions are shown in Eq. (12) below:

$$
f(never) = \begin{cases} 1, & \text{if } x = 0 \\ -500x+1, & \text{if } 0<x\leq 0.002 \\ 0, & \text{otherwise} \end{cases}
\qquad
f(seldom) = \begin{cases} 1, & \text{if } 0\leq x<0.002 \\ -500x+2, & \text{if } 0.002\leq x<0.004 \\ 0, & \text{otherwise} \end{cases}
$$

$$
f(rarely) = \begin{cases} 500x+0.5, & \text{if } 0\leq x<0.001 \\ 1, & \text{if } 0.001\leq x<0.003 \\ -500x+2.5, & \text{if } 0.003\leq x<0.005 \\ 0, & \text{otherwise} \end{cases}
\qquad
f(occasionally) = \begin{cases} 500x, & \text{if } 0\leq x<0.002 \\ 1, & \text{if } 0.002\leq x<0.004 \\ -500x+3, & \text{if } 0.004\leq x<0.006 \\ 0, & \text{otherwise} \end{cases}
$$

$$
f(infrequently) = \begin{cases} 500x-0.5, & \text{if } 0.001\leq x<0.003 \\ 1, & \text{if } 0.003\leq x<0.005 \\ -500x+3.5, & \text{if } 0.005\leq x<0.007 \\ 0, & \text{otherwise} \end{cases}
\qquad
f(sometimes) = \begin{cases} 500x-1, & \text{if } 0.002\leq x<0.004 \\ 1, & \text{if } 0.004\leq x<0.006 \\ -500x+4, & \text{if } 0.006\leq x<0.008 \\ 0, & \text{otherwise} \end{cases}
$$

$$
f(frequently) = \begin{cases} 500x-1.5, & \text{if } 0.003\leq x<0.005 \\ 1, & \text{if } 0.005\leq x<0.007 \\ -500x+4.5, & \text{if } 0.007\leq x<0.009 \\ 0, & \text{otherwise} \end{cases}
\qquad
f(often) = \begin{cases} 500x-2, & \text{if } 0.004\leq x<0.006 \\ 1, & \text{if } 0.006\leq x<0.008 \\ -500x+5, & \text{if } 0.008\leq x<0.01 \\ 0, & \text{otherwise} \end{cases}
$$

$$
f(regularly) = \begin{cases} 500x-2.5, & \text{if } 0.005\leq x<0.007 \\ 1, & \text{if } 0.007\leq x<0.01 \\ -500x+6, & \text{if } 0.01\leq x<0.012 \\ 0, & \text{otherwise} \end{cases}
\qquad
f(normally) = \begin{cases} 500x-4, & \text{if } 0.008\leq x<0.01 \\ 1, & \text{if } 0.01\leq x<0.03 \\ -500x+16, & \text{if } 0.03\leq x<0.032 \\ 0, & \text{otherwise} \end{cases}
$$

$$
f(usually) = \begin{cases} 500x-14, & \text{if } 0.028\leq x<0.3 \\ 1, & \text{if } 0.03\leq x<0.06 \\ -500x+31, & \text{if } 0.06\leq x<0.062 \\ 0, & \text{otherwise} \end{cases}
\qquad
f(generally) = \begin{cases} 500x-29, & \text{if } 0.058\leq x<0.06 \\ 1, & \text{if } 0.06\leq x<0.09 \\ -500x+46, & \text{if } 0.09\leq x<0.092 \\ 0, & \text{otherwise} \end{cases}
$$

$$
f(hardly\,ever) = \begin{cases} 500x-44, & \text{if } 0.088\leq x<0.09 \\ 1, & \text{if } 0.09\leq x<0.097 \\ -500x+49.5, & \text{if } 0.097\leq x<0.099 \\ 0, & \text{otherwise} \end{cases}
\qquad
f(always) = \begin{cases} 500x-48.5, & \text{if } 0.097\leq x<0.099 \\ 1, & \text{if } x\geq 0.099 \\ 0, & \text{otherwise} \end{cases}
$$

$$(12)$$

Functions of all describing words

When building the sentence, the word with the maximum membership degree is chosen because it will be the best match for the meaning that the sentence needs. Words with equal degrees can be used synonymously. The framework of the sentence is:

"Passengers are [c] travelling from (a) to (b)."

In this sentence, (a) is the starting point of the journey, while (b) is the destination point, and $[c]$ represents an adjective that will describe the frequency of that connection.

To gain deeper insight into the FCM, some specific connecting combinations are investigated. The first is the connection from the retirement home to the casino. The rate of passengers is 0.0018. Based on this value, the truth value is now calculated for every adjective. There are two words with a maximum truth value of 1, that is, seldom and rarely. These words can be used synonymously, which means that one of the two words can be selected randomly. The sentence to be built would then be:

*"Passengers are **rarely** travelling from the retirement home to the casino."*

The following ten sentences are built the same way:

1. Passengers are occasionally travel from suburb A to hotel C.
2. Passengers are rarely travel from hotel C to suburb A.
3. Passengers are seldom travel from the cemetery to the port.
4. Passengers are frequently travel from hotel A to the night club.
5. Passengers are rarely travel from the railway station to the convention center.
6. Passengers are seldom travel from the airport to the convention center.
7. Passengers are infrequently travel from company D to other companies.
8. Passengers are rarely travel from the theatre to the railway station.
9. Passengers are seldom travel from the hospital to the cemetery.
10. Passengers are sometimes travel from other companies to the hotel A.

There are some observations that can be made about these sentences. More passengers travel from suburb A to hotel C than vice versa (sentences 1 and 2). Travel to the convention center does not seem to be in high demand (sentences 5 and 6). In addition, hotel A must be in the center of the city because many connections from and to hotel A are highly ranked (sentences 4 and 10).

3.5   Future Application of the Model in Cognitive Cities

For the sake of simplicity, this use case considers only location nodes, but the model itself can be extended in many different directions, depending on the domain where it will be used. The model meets the requirements required to build a cognitive system that could be used in future cognitive cities. Because FCMs are already in use, mainly in domain-specific decision-making processes, extending the same model to applications that support humans is an obvious step.

By reducing large amounts of data to their essential components, users can gain insights and knowledge about a specific domain. These insights allow users to make better decisions because they have access to broader knowledge coverage—even for very complex problems. The system's ability to verbalize the hypotheses helps ensure more direct communication, so this model is most useful in applications where cognitive systems have direct contact with humans. If the model were extended with user-level features, it could adapt individually to every user. This individualization is an acknowledgement that users view the information relevant to their decision-making in a personal-contextual manner.

To summarize, this model can be a useful foundation for applications in cognitive cities whenever the personal knowledge of humans about a problem would benefit from amplification, enabling better decisions.

4 Conclusion

An FCM shows how concepts in a set and depicted as nodes interact with one another. The direction of the influence between nodes is shown with directed graphs, while the degree of interaction is defined by edge weights. The state of each node can be computed, and these states can then be used to identify influence patterns in the FCM. In parallel, learning algorithms can be applied to the edge weights so that parts of the cycles are excluded or highlighted. However, because all the results of an FCM are purely mathematical, interpreting FCM output is difficult. The solution is to add the RCT, which seeks to turn a restriction—i.e., a proposition in natural language—into computable content. The restriction is represented as a meaningful postulate that must be precisiated in a mathematical way. For this, an explanatory database provides the data source for the precisiation process.

To turn the mathematical content of an FCM into natural language, the process of the RCT must be reversed; in other words, a meaningful sentence must be created based on mathematical input. The FCM acts as the explanatory database, providing all the necessary information to create a proposition. However, the creation of a set of words that turns any value of the explanatory database into a sentence with a certain meaning is difficult because it strongly depends on the FCM's context. On the one hand, there must be enough granularity to not lose too much information; on the other hand, the complexity increases with every additional word in the descriptive set. It also must be considered that human perception of descriptive words is individual, and that the number of misperceived words is proportional to the number of different words. As a proof of concept, the use case described here shows that this idea can be realized. The engine that was introduced in this use case is already suitable for simple purposes.

Using the features of the proposed model, it is possible to build more complex environments that can be applicable and useful in cognitive computing. These more complex versions could therefore also be applicable for cognitive cities, especially where there is a need to logically connect concepts. Through the RCT, it is possible to give these connected concepts a computable meaning. There are still limitations in the methodologies underlying the proposed model. The approach used here, in which FCM output is converted into natural language with the aid of the RCT, requires interpretation of the FCM by the experts who build the FCM. They are in the lead in regard to defining the sentence framework, precisiating the set of words to be used, and specifying the associated membership functions. On the one hand, this presents an opportunity for the content to be made understandable to a larger community; on the other hand, it carries the risk that some information may be manipulated or withheld, either by mistake or on purpose.

To say this heretically, it means that end users must rely on the benevolence of the experts because the latter know which questions can be answered with the FCM. Consequently, it also means that this approach cannot be understood as data mining but rather must be considered data analysis because the RCT does not provide a means to find anything that is not instinctively suspected to be present already. However, the simulation presented in Sect. 2.3, where cycles can be recognized in an FCM, acts as a data mining tool. Therefore, it must be imperatively included in the interpretation tool. In his paper, Hagiwara [8] noted three major areas of improvement needed for common FCMs: (i) the proportionality of a relationship between two concepts, (ii) the lack of time delays; and (iii) the impossibility of representing multiple causality. The first point is especially interesting when trying to depict customer behavior in relation to price, which is usually not linear but elastic. It also shows that the true potential of this approach is still far from being fully exploited.

Just as cognitive cities are still a vision of the future, our model for a cognitive system remains in its infancy. Nevertheless, it already includes many features that are the foundation of a complete cognitive system. The next steps are to apply the model to types beyond just locations, with the goal of representing more complex datasets. With the added complexity, it would be possible to tackle business-related problems. Last, but not least, individualized configurations would be a great asset, so that a problem's context problem could be associated with an individual user's needs.

Cognitive systems can solve complex problems that cities will confront in the future. Therefore, it makes sense to investigate possible models today that best suit a cognitive system. IBM is currently on a path to introduce cognitive computing and cognitive systems. Cloud-based solutions allow such systems to process enormous amounts of data using different applications. Therefore, it would be useful to include this model in cloud-based applications.

References

1. Buzan, T.: Mind mapping. Kickstart your creativity and transform your life (Buzan bites). Pearson Education Limited (2006)
2. Chen, B., Cheng, H.H.: A review of the applications of agent technology in traffic and transportation systems. IEEE Trans. Intell. Transp. Syst. **11**(2), 485–497 (2010)
3. Dickerson, J.A., Kosko, B.: Virtual worlds as fuzzy cognitive maps. In: A. Chairperson (Chair) Virtual Reality Annual International Symposium, pp. 471–477. Symposium conducted at the meeting of IEEE, Location (1993)
4. Esser, S.K., et al.: Cognitive computing systems: Algorithms and applications for networks of neurosynaptic cores. In: Neural Networks (IJCNN), International Joint Conference, pp. 1–10. Location (2013)
5. European Parliament: Mapping smart cities in the EU. Retrieved from: http://www.smartcities.at/assets/Publikationen/Weitere-Publikationen-zum-Thema/mappingsmartcities.pdf (2014)
6. Hurwitz, J., Kaufman, H., Bowles, A.: Cognitive computing and big data analytics. Wiley, Indianapolis (2015)
7. Frase, K.: The promise of cognitive cities. http://citizenibm.com/2013/10/frase_cognitive_cities.html (2013). Retrieved 15 May 2015
8. Hagiwara, M.: Extended fuzzy cognitive maps. In: IEEE International Conference on Fuzzy Systems, pp. 795–801. Location (1992)

9. Kelly, J.E., Hamm, S.: Smart machines: IBM's Watson and the Era of cognitive computing. Columbia Business School Publishing (2013)
10. Kosko, B.: Fuzzy cognitive maps. Int. J. Man Mach. Stud. **24**(1), 65–75 (1986)
11. OECD Environmental Outlook to 2050. The Consequences of Inaction. OECD Publishing. http://dx.doi.org/10.1787/9789264122246-en (2012)
12. Özesmi, U., Özesmi, S.L.: Ecological models based on people's knowledge: a multi-step fuzzy cognitive mapping approach. Ecol. Model. **176**, 43–64 (2004)
13. Papageorgiou, E.I.: A review study of FCM applications during the last decade. In: Proceedings of IEEE International Conference of Fuzzy Systems, pp. 828–835. Taipei (2012)
14. Portman, E., Finger, M.: Smart cities—Ein Überblick. HMD Praxis der Wirtschaftsinformatik 304 (2015)
15. Portmann, E., Kaltenrieder, R., Pedrycz, W.: Knowledge representation through graphs. Journal **VV**(II), (2015)
16. Rodrigue, JP., Comtois, C., Slack, B.: The geography of transport systems. Routledge, New York (2013)
17. Schaffers, H., Komninos, N., Pallot, M., Trousse, B., Nilsson M., Oliveira, A.: Smart cities and the future Internet: Towards cooperation frameworks for open innovation. Future Internet Assembly, 2011, Achievements and Technological Promises, pp. 431–446. Heidelberg (2011)
18. Stach, W., Kurgan, L., Pedrycz, W., Reformat, M.: Genetic learning of fuzzy cognitive maps. Fuzzy Sets Syst. **153**, 371–401 (2005)
19. United Nations: World population prospects. Key findings and tabels. http://esa.un.org/ (2015)
20. Zadeh, L.A.: Toward a perception-based theory of probabilistic reasoning with imprecise probabilities. J. Stat. Plann. Infer. **105**, 233–264 (2012)
21. Zadeh, L.A.: Toward a restriction-centered theory of truth and meaning (RCT). Inf. Sci. **248**, 1–14 (2013)

Cognitive Cities: An Application for Nairobi

Text Message Participation of Slum Inhabitants to Improve Sanitary Facilities

Sara D'Onofrio, Noémie Zurlinden, Dominique Gadient and Edy Portmann

Abstract Population growth is always increasing, and thus the concept of smart and cognitive cities is becoming more important. Developed countries are aware of and working towards needed changes in city management. However, emerging countries require the optimization of their own city management. This chapter illustrates, based on a use case, how a city in an emerging country can quickly progress using the concept of smart and cognitive cities. Nairobi, the capital of Kenya, is chosen for the test case. More than half of the population of Nairobi lives in slums with poor sanitation, and many slum inhabitants often share a single toilet, so the proper functioning and reliable maintenance of toilets are crucial. For this purpose, an approach for processing text messages based on cognitive computing (using soft computing methods) is introduced. Slum inhabitants can inform the responsible center via text messages in cases when toilets are not functioning properly. Through cognitive computer systems, the responsible center can fix the problem in a quick and efficient way by sending repair workers to the area. Focusing on the slum of Kibera, an easy-to-handle approach for slum inhabitants is presented, which can make the city more efficient, sustainable and resilient (i.e., cognitive).

S. D'Onofrio (✉) · D. Gadient · E. Portmann
Institute of Information Systems, University of Bern, Bern, Switzerland
e-mail: sara.donofrio@iwi.unibe.ch

D. Gadient
e-mail: dominique.gadient@students.unibe.ch

E. Portmann
e-mail: edy.portmann@iwi.unibe.ch

N. Zurlinden
Swiss Institute for International Economics and Applied Economic Research,
University of St. Gallen, St. Gallen, Switzerland
e-mail: noemie.zurlinden@unisg.ch

© Springer International Publishing Switzerland 2016
E. Portmann and M. Finger (eds.), *Towards Cognitive Cities*,
Studies in Systems, Decision and Control 63, DOI 10.1007/978-3-319-33798-2_8

1 Introduction

Between 2014 and 2050, the global urban population is expected to grow by 63 % and the population is expected to grow by 32 % [50]. Furthermore, the major part of the global population is and will continue to be mostly located in developing countries [50], which includes some of the poorest people in the world [29]. Nearly 90 % of the global urban population growth will occur in African and Asian cities. These emerging countries will need to make room for an additional 1 billion citizens, while developed countries are expected to merely experience an increase of 68 million citizens [29].

Nairobi, as an example of a city in an emerging country, is and still will be faced with many challenges, such as inadequate infrastructure and air and water pollution. It is becoming increasingly difficult for this city to provide basic services and infrastructures to its citizens [36]. The concept of smart cities is thus not only promising and needed in developed countries but also in those that are emerging. This chapter aims to illustrate how the concepts of smart cities and cognitive cities can be applied to an emerging country by examining a use case for an African city to visualize how progress through cognitive computing can be presented.

Nairobi, the capital city of Kenya, has almost 3.5 million citizens [3] of whom 2 million live in slums (e.g., Kibera) [1]. The city faces severe problems including poor sanitation [65]. Slum inhabitants are particularly in need of the provision of better services, and therefore this use case presents an approach of how to make this possible.

One solution for Nairobi to cope with these challenges is for it to develop into a smart city. By applying cognitive computing, the city becomes more efficient as well as sustainable and resilient. In other words, a smart city can develop into a cognitive city by applying cognitive computing and can thus become more intelligent [39, 42].

Today, most city infrastructures are based on legacy systems, which prevent them from becoming cognitive. 92 % of the citizens of Nairobi possess mobile phones [37], but only a few people have access to computer systems. Thus, new and innovative approaches should first rely on already widespread (low-tech) information communication technologies, such as mobile phones (i.e., not smartphones). Eventually, if high-tech information communication technologies are more generally available, more advanced approaches can be developed.

Taking account of the widespread use of mobile phones in Kenya, the proposed approach enables communication between citizens and government institutions via text messages. More specifically, slum inhabitants can report problems with public toilets to the responsible center via text messages. Then, the messages will be processed automatically and allow for a convenient resolution of the problem [52]. Although the proposed approach relies on low-tech information communication technologies on the side of the users (i.e., the slum inhabitants), cognitive computing [22] is applied on the side of the server (i.e., the responsible center).

Focusing on low-tech solutions, this approach shows that, through the ever-increasing use of digital technologies, data progressively become more available and communication becomes more efficient. This enables the creation of adaptive systems to address complexity, providing an efficient method for tackling complex urban problems [34].

Furthermore, this approach shows that everyone can be an intelligent sensor [13]. By including slum inhabitants, good maintenance of the few existing toilets can be ensured. This shows that, in spite of facing severe challenges, cities can implement new digital possibilities [34], such as using citizens as sensors (cf. [13]). In other words, citizens are participants in designing the city and can have an impact through their participation [4]. This approach enables government institutions to straightforwardly provide services to citizens and enhance services in slums, thus helping Nairobi as a whole to become efficient (i.e., a smart city), sustainable (i.e., a learning city) and resilient (i.e., a cognitive city) [42].

Emerging countries should become more intelligent so that they have the ability to withstand challenges and consequences in the future (e.g., population growth). Although the authors pursue a design-science-oriented action design research method [44], this chapter does not present a ready-made application such as an instantiation of an artifact, but rather the chapter uses a case to demonstrate how emerging countries can become smarter through low-tech solutions with the help of soft computing methods. This is an outline of the current state of a work in process and is in line with transdisciplinary research [56].

The remainder of this chapter is structured as follows: Sect. 2 presents the theoretical background, Sect. 3 outlines the approach itself and Sect. 4 concludes this chapter.

2 Theoretical Background

This section provides an overview of the research on low-tech solutions in emerging countries and explains the necessary background of the presented approach. First, the concept of intelligent city development is introduced, followed by explanations of the use of complex adaptive systems, cognitive computing, and soft computing for cognitive cities.

2.1 Related Work

Ample literature shows that low-tech ICT may enhance the health of people living in rural areas. In particular, the literature investigates the potential of mobile phone use to improve living standards in Africa (see Table 1) as well as in other countries (see Table 2).

Table 1 Literature review I

Africa

Source	Approach		ICT	Method		Aim
Zurovac et al. [66]	As reminder	Kenya	Malaria Case Management	Quantitative	Cluster-randomized controlled trial	Measuring adherence to treatment guidelines
Pop-Eleches et al. [40]		Kenya	Short Message Service		Randomized controlled trial	Testing efficacy of SMS reminders on adherence to antiretroviral therapy
Mbuag-baw et al. [31]		Cameroon	Cameroon Mobile Phone SMS (CAMPS)		Single-centered randomized controlled single-blinded trial	Measuring medication adherence
Wakad-ha et al. [53]		Kenya	Open source system: Rapid SMS	Qualitative	Pilot study	Measuring feasibility of mobile phone-based system to remind mothers to vaccinate children
Hao et al. [14]	For better communication	Swaziland	LabPush System	Qualitative	Semi-structured and in-depth one on one interviews	Evaluating the LabPush system for sending/receiving laboratory results
Heimerl et al. [16]		Uganda	Voice Messages		Focus groups	Exploring the accessibility and value of voice messages
Vadhat et al. [51]	For better knowledge	Kenya	Mobile for Reproductive Health system	Mixed	Automatic logging of queries, demographic behavior change questions via SMS, telephone interviews	Evaluating acceptability, information access, and behavioral impact of providing contraception information via SMS

Table 2 Literature review II

Other countries

Source	Approach		ICT	Method		Aim
Chen et al. [7]	As reminder	Zhejiang (China)	SMS and Telephone	Quantitative	Randomized controlled trial with 3 groups: (1) control (no reminder), (2) SMS, (3) phone reminder group	Comparing efficacy of SMS and phone reminder to improve attendance rates at health promotion center
Martini da Costa et al. [30]		Brazil	HIV Alert System (HIVAS)		Randomized controlled trial	Measuring adherence satisfaction of SMS-based system for antiretroviral drug-based treatment regimens
Vogt et al. [52]	For better communication	Myanmar	Mobile Competence Network (MCN)	Design-oriented		Proposing an application (incl. system architecture and codebooks) accompanied by a use case
Anderson et al. [2]	For better knowledge	Central Asia	System *bus (with hardware device *box)	Mixed	(1) Ethnographic research, (2) Proof-of-concept system using GPS and SMS	Proposing system for improving access to transit information for bus riders
Fafchamps and Minten [11]		Maharashtra (India)	Reuters Market Light (RML)	Quantitative	Controlled randomized experiment	Assessing whether agricultural information distributed via SMS generates economic benefits

There has been some research on various issues; however, the majority of the research focuses on health issues in African countries. Some researchers have investigated low-tech solutions as reminders (i.e., medication or vaccination reminders), while others have explored the potential use of low-tech solutions for better communication or knowledge enhancement. These approaches mainly work through text messages. The applied approaches also make it possible to reach people in remote areas who would otherwise face poor health services.

Additionally, the approaches applied in countries outside of Africa seem to focus on health issues. There is one application that targets the transportation system.

To summarize the results, text/voice messages were used as a reminding device to improve communication or increase the knowledge of citizens, thus enhancing the living conditions in such countries. There are no known approaches target another pressing issue, namely sanitation facilities. This chapter introduces a novel aspect to the current research.

The previous literature review has provided an overview of the current approaches that attempt to improve the lives of poor people through technology. These approaches are first steps in the direction of smart and cognitive cities. The next section will focus on the concept of smart cities and on the transition to cognitive cities.

2.2 From Smart to Cognitive Cities

Even traditional (hard) computing techniques can enable efficient information processing and thus the development of smart cities [39, 41]. Smart cities can be realized by investing in communication infrastructure as well as in human and social capital. This can lead to higher standards of living, higher levels of sustainability [6] and smart systems and institutions (e.g., smart economy, people, governance, mobility, environment and living [15]). Humans can learn based on computer systems, and thus a city can become more sustainable by developing into a learning city [42].

To further improve city functioning (i.e., by making it more resilient), however, another level of learning and cognition is needed [42]. According to connectivism, people not only learn from their own experiences and perceptions but also from the knowledge of others, which calls for the assistance of computers to address the scope and complexity of available information [46]. Therefore, it is possible for all citizens to augment their individual cognition. Cognitive cities [34] rely on cognitive processes and systems that can learn based on past experiences, react to changes in their environments, and allow for efficiency and sustainability [22, 41, 42]. By empowering humans and cognitive systems to develop autonomously through the use of information communication technologies and by ensuring their proper cooperation, a city can then become resilient [42].

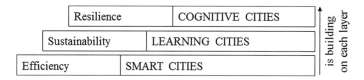

Fig. 1 City stack

Computer system resilience is defined as having resistant, absorptive and restorative capacities to prevent any possible hazards and to become more stable (i.e., equilibrium and maintenance of a predictable world) [12, 21, 38].

Illustrated in Fig. 1, cognitive cities are built based on learning cities, which in turn are built on smart cities. The next section discusses how cognitive cities can be developed through cognitive adaptive systems.

2.3   Development of Cognitive Cities Through Complex Adaptive Systems

A city can be defined as smart when the routing functions can be automated and more people have the ability to monitor, understand and analyze the data processed by the functions. People can then plan a city in a way to improve the standard of living in their city [4]. Technologies that provide computer systems with real-time awareness of reality and advanced analytics help citizens to make decisions that are more intelligent and thus foster the development of the city [8]. Complex adaptive systems provide one possibility for developing current cities into smart and further cognitive cities. These systems have a large number of components [20], so they consist of many subparts that interact with each other. The subparts are able to learn and adapt to new situations; complex adaptive systems address the complexity of urban systems resulting from the interaction between systems, governments, citizens, and ever-increasing databases [19]. Hence, appropriate solutions can be found based on the data and evidence.

Cognitive computing can be used to instantiate complex adaptive systems. This approach refers to the ability of automated computer systems to consciously, critically, logically and attentively handle information [54] and to learn and adjust to changes in a human-like way [22]. These computer systems are used to extend human intelligence [54] to improve the collaboration between humans and computer systems [26]. They acquire and analyze the right amount of information to make relevant decisions based on the given context [22]. To enable a straightforward interaction between humans and computer systems, which is necessary for building cognitive cities, cognitive computing is needed both to handle imprecise and complex data [22] and to address reality [26]. The next section introduces soft computing methods, which are related to cognitive computing and allow the system to handle imprecise data.

2.4   Introduction of Soft Computing Methods

Most data are imprecise (i.e., only available in natural language), so it is often difficult for computers to straightforwardly process inputs, and thus the use of soft computing methods should be considered. Fuzzy set theory [58] and fuzzy logic [59] are able to address language uncertainty and ambiguity. An element of a language set can belong to different classes, to a certain degree, because of fuzzy sets, and the element does not need to be accurately assigned to only one (crisp) class.

To obtain a full understanding of a situation (i.e., aggregation of possible (fuzzy) sets), granular computing should be used because this method allows for abstraction, generalization and clustering of information [57]. Granulation comprises an essential part of human cognition [63]. The ability to conceptualize reality in different levels of granularity is fundamental for human intelligence and flexibility, as it enables people to map the complexity of reality in easily understandable and comparable theories [18].

As granules in human reasoning are mostly fuzzy, fuzzy information granulation is needed [62]. In comparison to fuzzy sets, granules are not defined by membership functions but by clustering them together in classes by reason of similarity, proximity or functionality [61]. Fuzzy clustering, a part of fuzzy information granulation, is based on the idea that one element of a data set can belong to two or more classes with varying membership degrees. Thus, fuzzy clustering combines the abilities of fuzzy logic and granular computing, and this type of clustering allows the data to be simplified by a large degree. Fuzzy clustering can be done, for example, by mapping the complexity of real-time data in a triangle (see Fig. 2) [10].

A word can have different meanings, which in turn makes the interaction between humans and (computer) systems of a city more complicated [43]. Concretely, it is difficult for computers to process words and their meanings [33]. However, processing is made possible by the theory of computing with words [60].

Fig. 2 Fuzzy clustering

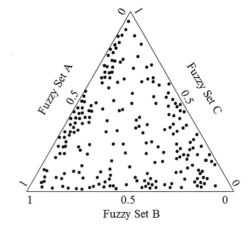

Fuzzy Set B

Computing with words permits automated reasoning, computing and decision-making based on linguistic variables (i.e., words) to occur [33]. In this context, information granulation is essential. A word is considered to be a label of a granule [60], and each granule has a value. The membership of a word to a granule can be defined by the similarity of this word to a considered value [28].

Furthermore, the restriction-centered theory (RCT) [64] should also be considered. A restriction R(X) defines the values a variable X can take. Most restrictions in human reasoning and cognition are often unprecisiated (i.e., described in natural language) and are thus fuzzy. Consequently, human perceptions are restrictions, and a natural language can be seen as a system of restrictions. To make the data computable, a precisiation is required [64]. In this context, the use of codebooks based on RCT [64] has the following advantages: by using simple codes, for example, it is easy for slum inhabitants to report any inconveniences with toilets. Reporting is quick through a short text message and thus promises to be an approach that will likely be used by most citizens. Furthermore, a common understanding enables data computation. This approach is explained in more detail in the next section.

3 The Approach

To address the complexity in the development of the approach, the authors apply an action design research approach [44] that is advanced through transdisciplinary research [56]. This combination is explained in the following section. Then, subsequent sections present the steps of transdisciplinary research, visualized by the chosen use case of Kibera. Finally, an evaluation concludes this subchapter.

3.1 Transdisciplinary Research for Enhancing Cities in Emerging Countries

Within action design research, knowledge can be generated from the given reality, meaning that the focal point is the collaboration between the research and the praxis to produce knowledge [25]. To have a structure, transdisciplinary research is used, meaning that there are three phases to the research: phase 1 is the analysis of the current state, phase 2 is the to-be analysis and phase 3 is the transformation (see Fig. 3).

In phase 1, collaboration with a team who is working locally (i.e., in Kibera) is planned; therefore, a possibility exists to provide real-time information about the current state of the local sanitary facilities and information communication technologies. The authors have an open dialogue with potential collaboration teams. Phase 2 describes the aim to be achieved through the research. In this case, the aim

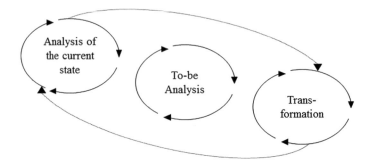

Fig. 3 Transdisciplinary research

is to develop a city in an emerging country into a cognitive city. Phase 3 presents the transformation from the current to the targeted state [35].

Furthermore, the authors use methods from design science research, which consist of designing a new notion to enhance the understanding of a problem and its possible solutions [17]. The next section explains step 1 of this approach.

3.2   Step 1: Analysis of the Current State

First, in order to clearly illustrate the advantages of the proposed approach, the low-tech solution will be explained step-by-step. Figure 4 provides an explanation of the procedure of the low-tech solution (i.e., sending and receiving text messages).

A text message is sent by a user (e.g., an inhabitant of Kibera) from a mobile phone; the message goes through a wireless link via a cell tower to a short message service center (SMSC). The SMSC stores the message and delivers it to the recipient (e.g., the responsible center) once it has a signal (i.e., reception and service) [5].

The use of GSM modems is a direct way to process a huge amount of incoming data. Text messages are a part of the Global System for Mobile Communications (GSM) and are available on any GSM mobile phone [52]. To permit interaction

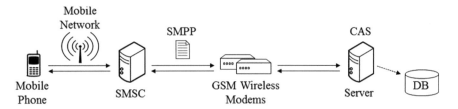

Fig. 4 Procedure of sending and receiving text messages

between SMSC and GSM, a Short Message Peer-to-Peer (SMPP) protocol (an open industry standard) is needed to simplify the integration of data applications in networks [5]. Text messages are then stored on a server (i.e., from the responsible center), such as in NoSQL-databases, and evaluated with complex adaptive systems to handle and visualize the (often unstructured) information [19]. NoSQL-databases (e.g., graph databases) have been chosen because they provide high performance, and have low latency, availability and the ability to address unstructured data [27].

Having analyzed the current state, the next step consists of analyzing what is targeted.

3.3   Step 2: To-Be-Analysis

The aim of this research is to develop computer systems that are more intelligent to help a city to become more efficient, sustainable and resilient [42]. Thereby, the city faces one big challenge: the natural language.

To allow for imprecision and uncertainty in the content of the text messages, fuzzy logic can be used. Specifically, as a first step, granular computing can be applied to find appropriate classes in the context of the sanitary facilities of Kibera. Possible classes are, for example, the different areas in the city.

Then, fuzzy clustering can be applied to group the text messages into classes according to their priority and toilet locations and problems (or any other applicable criteria). In this case, and for reasons of simplification (i.e., Occam's razor [47]), only three criteria are defined: location, problem and priority. The toilet location is required, the problem is required to address the repair issue, and the priority is needed to address the order in which the toilets should be repaired. Hence, data elements can be assigned to classes with varying membership degrees [24] (e.g., a toilet can partly belong to two areas and thus have potentially two different task forces).

Clustering is made possible by applying computing with words: codebooks serve as restrictions (i.e., shortcuts as codes) that permit the computation of the text message content (i.e., through precisiation based on restrictions [64]). Here, the following three codebooks are used: one for defining the toilet locations, one for reporting the problems, and one for specifying the priority.

The toilet location is defined by IDs. This means that every toilet possess a unique ID, which makes it possible to locate the toilet. According to the toilet-ID, the computer system can detect where the non-functioning toilet is located.

To describe the problem, a codebook is needed (see Table 3). To keep it simple, Table 3 is not exhaustive and only lists a limited number of cases.

Table 4 illustrates a priority codebook. According to Eisenhower Box [32], only the urgency dimension is integrated. Thus, only two priorities are presented below.

Many slum inhabitants do not have an advanced education and might not necessarily be fluent in English. Therefore, all codes are also stated in the other national language of Kenya, Swahili, and shortcuts based on these expressions can be used

Table 3 Codebook for problem definition

Shortcut English and meaning		Shortcut Swahili and meaning	
FDW	Flush does not work	KHK	Kuvuta haifanyi kazi
TIC	Toilet is clogged	CNM	Choo ni msongamano
TTB	Toilet is totally broken	CNKK	Choo ni kuvunjwa kabisa

Table 4 Codebook for priority specification

Shortcut English and meaning		Shortcut Swahili and meaning	
U	Urgent	H	Haraka
NU	Not urgent	SH	Si haraka

as well. It is important that illiterate people can also communicate with the central center via text message. One possible solution is weatherproof posters near the toilets, where pictures explain how to write a text message. Furthermore, these pictures visualize what the different problem cases (e.g., "flush does not work") look like and which priority code should be chosen. To prevent every inhabitant from stating that the problem is urgent, some rules such as "only if no toilet is working, then the priority *urgent* can be chosen." Of course, this rule should also be illustrated with pictures.

The different shortcuts (i.e., the two codebooks) are listed together with the ID-number of the toilet (e.g., 103 FDW U, 103 TIC U, 125 FDW NU etc.). This should help the inhabitant understand which shortcuts to use in which situations.

The content of the text messages can be computed and clustered. This allows the complex adaptive systems to recognize the problem type, the responsible area and sanitation team in Nairobi, and the text message language translation (i.e., from English to Swahili, or vice versa). Problems are then sent by the (cognitive) system to the responsible sanitation worker (via the same sending procedure as stated above) who is responsible for the coordination of all necessary tasks to fix the problem. After the problem is solved, the correspondent sanitation worker must send a text message with a specific code to the confirmation number. Then, the system automatically sends a text message to the inhabitants who reported the problematic toilet. In case of any trouble, the workers can use a different number to send a text message to the responsible center explaining their challenges.

The office workers in the responsible center have a relatively passive role: they coordinate operations and control all toilets in specific classes. The workers only actively react if and when challenges are reported; however, they respond to problematic or emergency situations and also control and report all collected data. In close dialog with the cognitive system, this enables the workers to understand patterns of malfunctioning (e.g., how many days pass on average until a specific toilet is broken again).

Computing with words attempts to extend fuzzy logic to optimize handling with complex adaptive systems by means of words [33] to mimic the way the human

Fig. 5 Slum inhabitant

Name	Tabita
Location	Kibera slum
Gender	Female
Age	20
Children	2
Occupation	Seller
Education	Primary school
Language	Swahili, Luo

brain works. A cognitive computer system can process text messages, classify the problems, respond easily, and, ultimately, solve the problems. Hence, a well-working interaction between humans and computer systems is one of the main characteristics of a cognitive (i.e., resilient) city. The next section visualizes, based on a use case, how the transformation from the current state to the targeted state should look.

3.4 Step 3: Transformation

To have a better comprehension of the use case, personas are first introduced and their roles are defined. Even if these personas only have fictive characteristics, they are based on information from real world scenarios [9, 23, 45, 55]. Then, the case is used to present how the approach should work. The evaluation of this use case in the next section concludes this subchapter.

3.4.1 Personas

In the use case of Kibera, three stakeholders are considered: slum inhabitants, sanitation workers and office workers. The first persona introduces the main stakeholder in this case: a slum inhabitant.

Two million Kenyan people live in slums [1], and the goal is to help slum inhabitants in emerging countries, so it is necessary to consider a slum inhabitant as a required party for this proposed approach. Figure 5 shows Tabita,[1] a 20-year-old mother of two children living in Kibera. She had primary education and is now working as a seller. She speaks Swahili and Luo. The use case will provide more

---

[1]Photo taken from http://www.photoworldwide.com/Exhibitions_and_Galleries/Pages/Akseli_ Gallen-Kallela%C2%B4s_Africa.html#80.

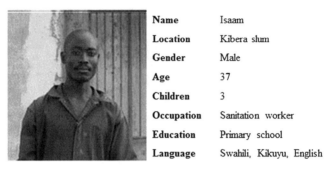

Name	Isaam
Location	Kibera slum
Gender	Male
Age	37
Children	3
Occupation	Sanitation worker
Education	Primary school
Language	Swahili, Kikuyu, English

Fig. 6 Sanitation worker

insight into the lives of slum inhabitants, as it is told from the perspective of Tabita. The next persona presents a Kiberian sanitation worker.

Other important participants in the proposed approach are the sanitation workers who repair the non-functioning toilets. In Fig. 6, Isaam,[2] a 37-year-old father, is presented. He also lives in Kibera and has a primary education. In comparison to Tabita, he speaks Kikuyu, Swahili and also English. He is a sanitation worker and also has a supervising and organizing role. Every day Isaam receives automatic text messages from the Nairobi Center telling him where in Kibera sanitation workers are required for repairing toilets. He delegates work to his team of 5 persons (most of the workers are very young) and is always available in the case of problems. When the non-functioning toilets are repaired, Isaam sends a text message to the Nairobi Center with the toilet ID-codes and the repaired status. He also may include other important information regarding the case if necessary to the Nairobi Center; for example, if any difficulties were encountered with repairing the toilets or with any of the slum inhabitants.

The last persona presents an administrator (i.e., an office worker) who is needed in the Nairobi Center to handle the entire processes.

To guarantee the success of the proposed approach, administrators are also needed. Figure 7 presents John,[3] a 28-year-old well-educated person who is working as an administrator in the Nairobi Center. John typically begins to investigate the reports automatically generated from the incoming and outgoing text messages and checks if any unfinished tasks remain. Then, he analyzes the data for predicting possible problems and tries to detect patterns in the incoming text messages, such as identifying the most frequently occurring problems or which toilets/areas are most prone to malfunction.

Thus, the most important stakeholders in the proposed approach have been introduced; the next section discusses the use case.

---

[2]Photo taken from http://www.metroafrican.org/category/the-arts/page/3/.

[3]Photo taken from http://www.jambonewspot.com/nairobi-man-unmasked/.

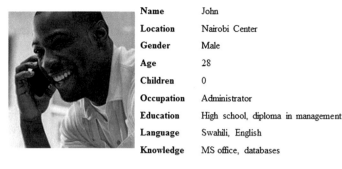

Name	John
Location	Nairobi Center
Gender	Male
Age	28
Children	0
Occupation	Administrator
Education	High school, diploma in management
Language	Swahili, English
Knowledge	MS office, databases

Fig. 7 Administrator

3.4.2 Kibera Use Case

This use case visualizes how the proposed approach helps improve the sanitary facilities of Kibera by including different parties (i.e., slum inhabitant, sanitation worker and administrator).

Tabita notices inoperative toilets near her workplace. Remembering a government initiative that aims to improve the sanitary situation in slums with the help of the inhabitants themselves, she looks for the poster hanging beside the toilets. On this poster, possible toilet problems are listed, both in words (English and Swahili) and in pictures, together with the respective problem codes including the toilet-ID, the urgency rule and the phone number of the Nairobi Center (i.e., the institution responsible for the maintenance of the toilets in the area). After having identified the problems (e.g., one toilet is clogged and another is completely broken), the urgency (e.g., all toilets are non-functioning, thus, it is urgent) and the respective Swahili shortcuts, Tabita sends a text message with the codes and the toilet-IDs (i.e., 104 CNM H, 105 CNKK H) to the Nairobi Center. This is the first step in the process and has to be conducted by Tabita to initiate the proposed approach. The second step occurs within the system when the text message is processed and automatically assigned to the respective class(es) by granular computing and fuzzy clustering. Thereby, the system automatically calculates the most efficient way to repair the broken toilets (e.g., how many workers for an area are needed and in which order the toilets should be repaired). Based on this knowledge, the third step includes when the cognitive computer system automatically passes the message and instructions to Isaam, the responsible sanitation worker (i.e., the coordinator) for the area where the toilets are located. At the same time, Tabita receives a response (i.e., "Thank you for your message. We will be there as soon as possible." in Swahili), automatically sent by the system. The fourth step consists of repairing the toilets. The most efficient way to tackle the problem has already been calculated by the system at this point, so Isaam only has to delegate the tasks. After the toilet is repaired, Isaam sends a text message to the Nairobi Center to confirm the tasks are completed. This leads to the final step in which the system automatically sends a confirmation message (i.e., "The toilet works again. Thank you for your patience." in Swahili) to Tabita.

3.5 Evaluation

The transdisciplinary approach splits the approach into three parts: the current state, the targeted state and the transformation. This use case shows how the proposed approach can be applied to a very specific problem in a specific area (i.e., the current state) to improve the standard of living (i.e., to transform it into the targeted state). This case only considers two inoperative toilets in Kibera, so the use case presents a simplified example. Normally, the system is required to process many different messages coming from different areas. The various problems are automatically processed, and the best possible solution method is calculated by the system. To allow the system to cope with imprecise data (i.e., text messages), fuzzy clustering and computing with words are applied. The key advantage of this approach is that slum inhabitants need only a mobile phone to communicate with the responsible center. Thus, the method provides a very convenient and simple method of communication, and it increases the local standards of living.

The Nairobi Center is the most important part in this approach. Even if the individuals working at the Nairobi Center do not directly interact with the slum inhabitants or with the sanitation workers, the (cognitive) computer systems do. The entire organization of the sanitary facilities occurs in the Nairobi Center. The incoming text messages must be assigned correctly to the classes that are defined by the situation with granular computing. Then, the messages are further processed to initiate the toilet reparation; therefore, the conversion process result of the incoming text messages is crucial.

In addition, the Nairobi Center works as an intermediate part in this approach. A slum inhabitant is never required to write a sanitation worker directly or vice versa. The Nairobi Center manages the entire communication channel and helps to optimize the task assignments. Thanks to the (cognitive) computer systems, the administrators can analyze the data and potentially detect patterns to better plan city development.

Every involved participant can focus on his or her own task with this method. Slum inhabitants are only required to write a short text message to have their problem solved, sanitation workers know the precise location to visit to repair a toilet, and the administrators manage the entire procedure and analyze the data to predict future cases. With the assistance of (cognitive) computer systems, Nairobi can become resilient when the communication between the stakeholders of this slum (i.e., part of the city) is fostered.

4 Conclusions and Outlook

This chapter presents an approach using cognitive computing and soft computing methods to enhance the living standards, in this case, in Kibera, the largest slum in Nairobi. Mobile phones (i.e., low-tech ICT) have potential to improve living conditions in emerging countries by providing the possibility to send and receive text messages. The application of complex adaptive systems in combination with soft computing methods allows cognitive cities to be developed.

Most people in Nairobi have at least one mobile phone, so this approach reaches the majority of the local population. By taking advantage of cognitive computing and soft computing methods, applications that are based on low-tech solutions have the power to considerably increase the quality life for the slum inhabitants. As more urban data becomes available, this approach can also calculate approximately how many toilets will be needed and further improve the sanitary situation, based on predicted population growth rates for the coming years (as in [49]). This enables the government to provide needy areas with more toilets in an effective way. Based on the data collected, the government can gain meaningful insights into slum systems and create sustainable, low-cost solutions to improve the lives of the citizens who particularly need it.

The authors are convinced that technological progress in emerging countries should begin with low-cost solutions. Then, after successful implementation, the solutions may be further expanded (see [39]). Moreover, the proposed approach can also be applied to other problems, such as water management or traffic and to cities in other emerging countries [39]. There are many other implementation possibilities to include inhabitants in city management beyond mere communication via text messages (e.g., [48]).

This approach is in the early stages. Further research should more carefully investigate the requirements of slum inhabitants in Nairobi to determine further problematic areas and to optimize the proposed approach using cognitive computing and soft computing methods. The authors are still searching for a collaboration partner located in Nairobi who is working on similar issues (preferably in Kibera). By eliciting the requirements of the slum inhabitants, the personas presented in this chapter can be adapted to correspond to reality as closely as possible.

References

1. Amnesty International: Slums in Kenia – Auf engstem Raum. http://www.amnesty.de/mit-menschenrechten-gegen-armut/wohnen-in-wuerde/auf-engstem-raum (2009). Accessed 13 Aug 2015
2. Anderson, R.E., Poon, A., Lustig, C., Brunette, W., Borriello, G., Kolko, B.E.: Building a transportation information system using only GPS and basic SMS infrastructure. In: Proceedings of the 3rd International Conference on Information and Communication Technologies and Development, pp. 233–242 (2009)
3. Auswärtiges Amt: Kenia. http://www.auswaertiges-amt.de/DE/Aussenpolitik/Laender/Laenderinfos/01-Nodes_Uebersichtsseiten/Kenia_node.html (2015). Accessed 13 Aug 2015

4. Batty, M., Axhausen, K.W., Giannotti, F., Pozdnoukhov, A., Bazzani, A., Wachowicz, M., Ouzonis, G., Portugali, Y.: Smart cities of the future. Eur. Phys. J. Special Topics **214**, 481–518 (2012)
5. Brown, J., Shipman, B., Vetter, R.: SMS: the short message service. Computer **40**(12), 106–110 (2007)
6. Caragliu, A., Del Bo, C., Nijkamp, P.: Smart cities in Europe. In: 3rd Central European Conference in Regional Science—CERS, 2009
7. Chen, Z., Fang, L., Chen, L., Dai, H.: Comparison of an SMS text messaging and phone reminder to improve attendance at a health promotion center: a randomized controlled trial. J. Zhejiang Univ. Sci. B **9**(1), 34–38 (2008)
8. Chourabi, H., Nam, T., Walker, S., Gil-Garcia, J.R., Mellouli, S., Nahon, K., Pardo, T.A., Scholl, H.J.: Understanding Smart Cities: An Integrative Framework. In: 45th Hawaii International Conference on System Sciences, pp. 2289–2297 (2012)
9. Desgroppes, A., Taupin, S.: Kibera: the biggest slum in Africa? Les Cahiers de l'Afrique de l'Est **44**, 23–34 (2011)
10. D'Onofrio, S., Portmann, E.: Von Fuzzy-Sets zu Computing-with-Words. Informatik-Spektrum **38**, 1–7 (2015)
11. Fafchamps, M., Minten, B.: Impact of SMS-based agricultural information on indian farmers. World Bank Econ. Rev. **26**(3), 383–414 (2012)
12. Ganin, A., Massaro, E., Gutfraind, A., Steen, N., Keisler, J.M., Kott, A., Mangoubi, R., Linkov, I.: Operational resilience: concepts, design and analysis. Nature Scientific Reports (2015) (under review)
13. Goodchild, M.F.: Citizens as Sensors: Web 2.0 and the volunteering of geographic information. GeoFocus (Editorial) **7**, 8–10 (2007)
14. Hao, W., Hsu, Y., Chen, K., Li, H., Iqbal, U., Nguyen, P., Huang, C., Yang, H., Lee, P., Li, M., Hlatshwayo, S., Li, Y., Jian, W.: LabPush: a pilot study of providing remote clinics with laboratory results via short message service (SMS) in Swaziland, Africa—a qualitative study. Comput. Methods Programs Biomed. **118**, 77–83 (2015)
15. Haubensak, O.: Smart cities and internet of things. In: Business Aspects of the Internet of Things, Seminar of Advanced Topics, ETH Zurich, pp. 33–39 (2011)
16. Heimerl, K., Honicky, R., Brewer, E., Parikh, T.: Message phone: a user study and analysis of asynchronous messaging in rural uganda. In: SOSP Workshop on Networked Systems for Developing Regions (NSDR), California (2009)
17. Hevner, A.R., March, S.T., Park, J., Ram, S.: Design science in information systems research. MIS Q. **28**(1), 75–105 (2004)
18. Hobbs, J.R.: Granularity. In: Proceedings of International Joint Conference on Artificial Intelligence (IJCAI), Los Angeles, CA, pp. 432–435 (1985)
19. Holland, J.H.: Complex adaptive systems. Daedalus **121**(1), 17–30 (1992)
20. Holland, J.H.: Studying complex adaptive systems. J. Syst. Sci. Complexity **19**, 1–8 (2006)
21. Holling, C.S.: Resilience and stability of ecological systems. Annu. Rev. Ecol. Syst. **4**(1), 1–23 (1973)
22. Hurwitz, J.S., Kaufman, M., Bowles, A.: Cognitive Computing and Big Data Analytics. John Wiley and Sons Inc, Hoboken, New Jersey (2015)
23. Integra: Life in Kibera slum. http://www.integra.sk/en/download/news/127-life-in-kibera-slum (2012). Accessed 27 Aug 2015
24. Jafar, O.A.M., Sivakumar, R.A.: Comparative Study of hard and fuzzy data clustering algorithms with cluster validity indices. In: Proceedings of International Conference on Emerging Research in Computing, Information, Communication and Applications (ERCICA 2013), pp. 775–782. Elsevier Publications (2013)
25. Järvinen, P.: Action research is similar to design science. Qual. Quant. **41**(1), 37–54 (2007)
26. Kelly III, J.E., Hamm, S.: Smart Machines, IBM's Watson and the Era of Cognitive Computing. Columbia University Press, New York, Chichester, West Sussex (2013)

27. Kumar, R., Gupta, N., Charu, S., Jangir, S.K.: Manage Big Data through NewSQL. In: National Conference on Innovation in Wireless Communication and Networking Technology, Association with THE INSTITUTION OF ENGINEERS INDIA (2014)
28. Lawry, J.: A methodology for computing with words. Int. J. Approximate Reasoning **28**, 51–89 (2001)
29. LSECities: Cities and the New Climate Economy: the transformative role of global urban growth. The New Climate Economy—The Global Commission on the Economy and Climate, pp. 1–70 (2014)
30. Martini da Costa, T., Peres Barbosa, B.J., Gomes e Costa, D.A., Sigulem, D., de Fatima Marin, H., Castelo Filho, A., Torres Pisa, I.: Results of a randomized controlled trial to assess the effects of a mobile SMS-based intervention on treatment adherence in HIV/AIDS-infected Brazilian women and impressions and satisfaction with respect to incoming messages. Int. J. Med .Inf. **81**(4), 257–269 (2012)
31. Mbuagbaw, L., Thabane, L., Ongolo-Zogo, P., Lester, R.T., Mills, E., Volmink, J., Yondo, D., José Essi, M., Bonono-Momnougui, R.C., Mba, R., Ndongo, J.S., Nkoa, F.C., Atangana Ondoa, H.: The cameroon mobile phone sms (CAMPS) tiral: a protocol for a randomized controlled trial of mobile phone text messaging versus usual care for improving adherence to highly active anti-retroviral therapy. BioMed Central **12**(5), 1–8 (2011)
32. McKay, B., McKay, K.: The Eisenhower decision matrix: how to distinguish between urgent and important tasks and make real progress in your life. http://www.artofmanliness.com/2013/10/23/eisenhower-decision-matrix/ (2013). Accessed 03 Dec 2015
33. Mendel, J.M., Zadeh, L.A., Trillas, E., Yager, R., Lawry, J., Hagas, H., Guadarrama, S.: What computing with words means to me. IEEE Comput. Intell. Mag. **5**(1), 20–26 (2010)
34. Mostashari, A., Arnold, F., Mansouri, M., Finger, M.: Cognitive cities and intelligent urban governance. Network Ind. Q. **13**(3), 4–7 (2011)
35. Moyser, R.: Planning for smart cities in the UK. http://www.burohappold.com/blog/post/planning-for-smart-cities-in-the-uk-2179/ (2013). Accessed: 02 May 2016
36. Mundia, C.N., Murayama, Y.: Modeling spatial processes of urban growth in African cities: a case study of Nairobi City. Urban Geogr. **31**(2), 259–272 (2010)
37. NCSS2 Final Report 2012: Population and health dynamics in Nairobi's informal settlements. http://aphrc.org/wp-content/uploads/2014/08/NCSS2-FINAL-Report.pdf (2012). Accessed 11 Aug 2015
38. Ouyang, M., Dueñas-Osorio, L., Min, X.: A three-stage resilience analysis framework for urban infrastructure systems. Struct. Saf. **36–37**, 23–31 (2012)
39. Picon, A.: Smart cities: a spatialised intelligence. John Wiley and Sons Inc., Hoboken, New Jersey (2015)
40. Pop-Eleches, C., Thirumurthy, H., Habyarimana, J.P., Zivin, J.G., Goldstein, M.P., de Walque, D., MacKeen, L., Haberer, J., Kimaiyo, S., Sidle, J., Ngare, D., Bangsberg, D.R.: Mobile phone technologies improve adherence to antiretroviral treatment in a resource-limited setting: a randomized controlled trial of text message reminders. AIDS **25**(6), 825–834 (2011)
41. Portmann, E., Finger, M.: Smart Cities – Ein Überblick! HMD Praxis der Wirtschaftsinformatik, pp. 1–12 (2015)
42. Portmann, E., Finger, M.: What are cognitive cities? In: Towards Cognitive Cities: Advances in Cognitive Computing and its Applications to the Governance of large Urban Systems. Springer International Publishing (2016) (In press)
43. Rappaport, W.J.: What did you mean by that? Misunderstanding, negotiation, and syntactic semantics. Mind. Mach. **13**, 397–427 (2003)
44. Sein, M.K., Henfridsson, O., Purao, S., Rossi, M., Lindgren, R.: Action design research. MIS Q. **35**(1), 37–56 (2011)
45. Shofco: Kibera. http://www.shofco.org/locations/kibera (2015). Accessed 27 Aug 2015
46. Siemens, G.: Connectivism: a learning theory for the digital age. Int. J. Instr. Technol. Distance Learn. **2**(1), 3–10 (2005)

47. Soegaard, M: Occam's Razor: The simplest solution is always the best. Interaction Design Foundation. https://www.interaction-design.org/literature/article/occam-s-razor-the-simplest-solution-is-always-the-best (2015). Accessed 03 Dec 2015
48. Terán, L., Meier, A.: SmartParticipation—a fuzzy-based platform for stimulating citizens' participation. Int. J. Infonomics (IJI) 4(3/4), 501–512 (2011)
49. UN-HABITAT: The State of the World's Cities Report 2006/2007. United Nations Human Settlements Programme (2006)
50. United Nations University:. Smart cities for sustainable development. http://egov.unu.edu/research/smart-cities-for-sustainable-development.html#outline (2015). Accessed 01 Dec 2015
51. Vadhat, H.L., L'Engle, K.L., Plourde, K.F., Magaria, L., Olawo, A.: There are some questions you may not ask in a clinic: providing contraception information to young people in Kenya using SMS. Int. J. Gynecol. Obstet. 123, e2–e6 (2013)
52. Vogt, J., Martin, E., Portmann, E., Mahmud, N.: Towards an SMS-based social network for health workers in rural areas in Myanmar. In: 3rd International Conference on Informatics, Electronics and Vision, Dhaka, Bangladesh (2014)
53. Wakadha, H., Chandir, S., Were, E.V., Rubin, A., Obor, D., Levine, O.S., Gibson, D.G., Odhiambo, F., Laserson, K.F., Feikin, D.R.: The feasibility of using mobile-phone based SMS reminders and conditional cash transfers to improve timely immunization in rural Kenya. Vaccine 31, 987–993 (2013)
54. Wang, Y.: Cognitive informatics: towards the future generation computers that think and feel. In: Yao, Y.Y., Shi, Z.Z., Wang, Y., Kinsner, W. (eds.) Proceedings of the 5th IEEE International Conference on Cognitive Informatics (ICCI'06), Beijing, China, pp. 3–7 (2006)
55. WASH United: Stigmatization in the realisation of the right to water and sanitation. http://www.wash-united.org/files/wash-united/resources/Stigmatization%20Report%20Final.pdf (2012). Accessed 30 Aug 2015
56. Wickson, F., Carew, A.L., Russell, A.W.: Transdisciplinary research: characteristics, quandaries and quality. Futures 38(9), 1046–1059 (2006)
57. Yao, Y.: Perspectives of granular computing. Proceedings of the IEE 2005 Conference on Granular Computing (GrC05), Beijing, vol. 1, pp. 85–90 (2005)
58. Zadeh, L.A.: Fuzzy sets. Inf. Control 8(3), 338–353 (1965)
59. Zadeh, L.A.: Fuzzy logic. Computer 21(4), 83–93 (1988)
60. Zadeh, L.A.: Fuzzy Logic = Computing with Words. IEEE Trans. Fuzzy Syst. 4(2), 103–111 (1996)
61. Zadeh, L.A.: The key role of information granulation and fuzzy logic in human reasoning, concept formulation and computing with words. In: Proceedings of IEEE 5th International Fuzzy System (1996)
62. Zadeh, L.A.: Some reflections on soft computing, granular computing and their roles in the conception, design and utilization of information/intelligent systems. Soft. Comput. 2, 23–25 (1998)
63. Zadeh, L.A.: From computing with numbers to computing with words—from manipulation of measurements to manipulation of perceptions. Ann. N. Y. Acad. Sci. 929(1), 221–252 (2001)
64. Zadeh, L.A.: Toward a restriction-centered theory of truth and meaning (RCT). Inf. Sci. 248, 1–14 (2013)
65. Zulu, E.M., Konseiga, A., Darteh, E., Mberu, B.: Migration and urbanization of poverty in sub-Saharan Africa: the case of Nairobi city, Kenya. In: Presented at the 2006 Annual Meeting of the Population Association of America Los Angeles California [Unpublished] (2006)
66. Zurovac, D., Sudoi, R.K., Akhwale, W.S., Ndiritu, M., Hamer, D.H., Rowe, A.K., Snow, R. W.: The effect of mobile phone text-message reminders on Kenyan health workers' adherence to malaria treatment guidelines: a cluster randomised trial. Lancet 378, 795–803 (2011)

Innovative Urban Governance: A Game Oriented Approach to Influencing Energy Behavior

Mo Mansouri and Nalan Ilyda Karaca

Abstract We are moving fast toward a smart urban era in which all components of our dominating urban life need to be not only sensing and smart but also learning and cognitive. This includes all stakeholders of the extensive network of urban systems and their governing body structure all across societal sectors. From this perspective, and considering the two key concepts of sensing and learning, the focus of the future efforts should be on education and collaboration among stakeholders to increase the entire system's reliability and effectiveness. There's also the obvious factor of infrastructure for machine-based sensing, which is becoming more available through the use of smart devices as well as venues for connectivity, which is fortunately possible by advancements and availability of social networking technology nowadays. Moreover, urban stakeholders' collaboration is also known to foster sustainable development, when accompanies a detailed plan for influencing behavioral patterns in city networks. Creating a dynamic methodology for policymaking through quantitative and computational governing mechanisms could be considered as a solution. The premise of this chapter is that innovative urban governance can effectively influence the desired impact and behavior through using technological tools such as city sensors and smart devices. Since administration of social games in smart urban environments are known as one of such governing mechanism, we have applied the theories of interactive collaboration to a simple yet effective game, which involves citizens of an isolated environment with dynamic adjustment of their behavior with regards to energy consumption. The presented case of Hoboken in this chapter is focused on evaluation of behavioral change among residents of Stevens Institute of Technology, given the right collective information and provided the sustainable incentive structure. As a part of this research, the results of the experiment with on campus residents were analyzed against similar information collected from citizens of Hoboken. The results of our

M. Mansouri (✉)
School of Systems and Enterprises, Stevens Institute of Technology,
Castle Point on Hudson, Hoboken, NJ 07030-5991, USA
e-mail: mo.mansouri@stevens.edu

N.I. Karaca
Information Technology Department, Credit Suisse (USA) LLC, New York, NY, USA

© Springer International Publishing Switzerland 2016
E. Portmann and M. Finger (eds.), *Towards Cognitive Cities*,
Studies in Systems, Decision and Control 63, DOI 10.1007/978-3-319-33798-2_9

research supports the hypothesis that people will choose a sustainable alternative when given the right information and provided with incentives to do so.

1 Introduction

The world is developing continuously, moving on its way to become a global village. We are living in a new era of technological prevalence and data abundance. The concepts of smart technologies have become an important part of our lives just to help us keeping up with requirements of 21st century. Many of these concepts, designed systems, and models are not so new. However, we had never faced with such a rapid move in urbanization and systems integration before throughout history. The high frequency of streaming continuous and never-ending new collections of data, as one of the major consequences of these changes, requires an update on the previously adopted approaches and models, almost in any field of science and technology. Concurrently, the integration around the world is happening in a rapid rate pushing for creation of transnational common aspects and technologies. It consequently creates an ever-changing environment, which demands for development of new models of governance, overlooking new dynamics in social interactions and offering enhanced solutions.

Revolutionary changes coming to arena of data collecting processes through smart sensing in urban systems of our time has drastically altered ways through which cities are designed, planed and viewed. It has also brought about new paradigms to the horizon of planning and management of cities that are elevated to the level of cognitive entities. Many methodic approaches have been also introduced to the field. "The new science of cities" for instance, is a concept that claims to find or develop novel applications of scientific methods and computational reasoning in urban design, development and management [1]. In the same manner, advancement of technology and availability of data can make a breakthrough in governing mechanisms applied to complex urban environment. These new governing mechanisms should be adopted from the results of experiments done in systems science research areas as well as in management of complex systems or similar concepts known as "system of systems" [2, 3].

These newly developed approaches can be institutionalized and applied to a concept previously introduced to the literature, known as "cognitive cities" [4] defined to be the ultimate manifestation of networked environments in which autonomic groups of urban systems coexist, compete, and collaborate to reach a higher collective purpose [5]. This necessity is partially affected by the new movement in the world toward designing, engineering and building smart cities in which all the stakeholders are provided with the effective options for adopting better choices. A smart city is a launching platform toward obtaining and institutionalizing cognitive capabilities within the fibers of urban systems. Stepping up the cognitive ladder can presumably bring about a revolution in the way every single citizen makes decisions in urban environments. In a higher level, it involves the complexity of interactions among all social stakeholders including governing bodies, law-makers (including and not limited to policymakers) as well as private and non-profit sectors. The efforts for capturing these complexities, beyond any shadow of doubt, are transforming how cities of future are supposed to be governed.

2 A Paradigm Shift in Governance

Embedding technologically updated governing approaches in urban systems require a paradigm shift in the realm of urban governance. Accessibility of data to the right societal entities demands for a stakeholder-based approach to the problem of governance and like any other similar perspective, such new governing model is only effective when applied in a systemic context through which immerging school of thoughts in governing urban systems evolve. The intensity of interactions caused by power structure in society from one side and importance of proposing a holistic solution for the situation at hand from the other, limit the effectiveness of governance. To address this shortcoming, governing approaches should consider a balanced interaction between bottom-up evolving processes and top-down organizing enforcements. This is in contrast with a lot of classic governance schools of thoughts, which consider a controlling role for governing bodies through using law-based regulations and leveraging coercion in parallel.

The new interactive and recursive paradigm, such as the one suggested by pioneers of the new sciences for cities [6], should bring the entirety of future urban systems to the level of an adaptive organism that is in constant interaction with its environment to sense social reactions and implement necessary changes to its governing mechanisms. The interactive dynamics of all actors in urban environment supposedly animates the forces of change and growth, when aligned meticulously with planning of critical facilities and transactional movements within the urban systems as is argued in the literature [7]. To fathom all aspects of this new outlook to urban governance, it is necessary to have an understanding of technological advancements and available sensing capabilities. Technology has a direct impact on our capability to build smart, or even optimistically, cognitive cities, which beyond sensing and disseminating flows of information, are also capable of learning collectively throughout the time and evolve accordingly to adapt the emergencies of their environment.

The next step is to develop an understanding on the evolutions of analytical methods and computational capabilities. The epistemological shift in data analysis, especially considering the amount of data that is available and presented to us using the new technologies will allegedly change the previously used transactional equilibrium [1]. The new paradigm, therefore, is expectedly inclusive of systems concepts and sciences by appreciating agent based modeling, simulations and analysis, and applying relevant newly developed scientific models such as applications of chaos and complexity theories in pattern recognition, social network theory, and so forth. Ultimately, the following step is to include and imply such newly developed discussions and visions toward changing behavioral patterns of citizens at large, using recursive models, which brings back corrective policies, regulations and policies to the existing governing system. Despite their level of individual power, citizens are the main stakeholders of cities and their collective actions will equip urban systems with variety of smart solutions in future.

In fact, citizens are an important part of cognitive capabilities in the cities of future and their solutions are going to be the key factors in keeping the balance in all aspects of life. They will be the balancing forces from "bottom" and they will be able to impact cities profoundly through enhancing economic and ecological sustainability as well as decisions they make in variety of fields including energy, water, transportation, communication, education, and healthcare. In better words, cognitive cities are the ones that can learn over time and possibly influence the behavior of their citizens through structural and systemic directions. That type of learning organisms could be considered as cognitive cities in which the process of learning is established and hence, they have the capacity of adopting effective and efficient policies through intelligent governing methods [4].

3   Governing Behavior Through Social Networks

There are many other concepts and methodologies that are adopted from social sciences perspectives that along with computational methods and models, give life to the aforementioned new paradigm of governance. This way, quantitative solutions such as mathematical modeling, simulation, optimization methods, etc. will get combined with qualitative approaches of political sciences, sociology and social psychology to empower the governance structure in design, engineering, development, and management of cities. Moreover, considering the importance and level of connectivity that nowadays social, physical, and cyber-physical networks have created around us [8], such new paradigm of governance could also influence the existing social structures and particularly collective behaviors of the citizens.

The combinatorial scientific approach for city governance is being done through mathematical modeling in data sciences, that is, extensive analysis of existing data, coupling them with technological advancements and innovative new approaches. The constructed toolset will then be used in translating collected data into design, engineering, development and governing guidelines. The social network and societal side of the problem, on the other hand, is being done through variety of ways that exist for social sensing, including but not limited to: online social network analysis and application oriented smart phone data collection methods. Analytics of social network is the subject of research in other fields of computational sciences and is also done by online social network enterprises such as Facebook and Twitter as well as others, which collect information using software-installed cookies.

As for the cognitive cities, it is first necessary to collect online data and understand the structure of social networks based on collected information. Only then, we will be capable of applying other quantitative methods such as system dynamics as well as agent based modeling and simulation methodologies to provide policymaking entities with computational ways for evaluating policies and their impacts on social networks (as a micro model of society at large) before enforcing them. Showing applications of social network analysis requires access to a huge amount of data and is out of this chapter's scope. The presented computational frameworks, paradigms, and models in this chapter however, are solidified by a case study on mechanisms for incentivizing behavioral change in regards to sustainable energy consumption in an isolated subset of a small size American city.

Governing energy behavior is becoming one of the most important issues among the portfolio of possible toolsets for urban behavior management, considering the exacerbation of energy cost. Although other aspects of governance for urban systems such as water, waste, education, communication, health, etc. could be used too, governing energy behavior as a module has the potential to be scaled up to governing others. Moreover, institutionalization of sustainable behavior is crucial for flourishing and growth of any urban system, both in terms of ethics and economics. Energy behavior has a pivotal role in this effort toward sustainability and based on such reasoning the presented case in this chapter is chosen. The case combines qualitative and quantitative methodologies to understand the mechanisms for influencing change in a social behavioral context with respect to energy consumption in Hoboken, as small town on the banks of Hudson River, facing New York City from the state of New Jersey (NJ) in the United States of America (USA). The case study is conducted and documented within the context of an award winning academic thesis as a partial requirement to obtain a Master of Science degree at the School of Systems and Enterprises (SSE) in Stevens Institute of Technology (SIT) [9]. The case is built upon information gathered from hundreds of surveys and interviews and advanced to a computational systems dynamics model for further investigation into policies that incentivize citizens to adopt sustainable behavior.

4 Current Concerns of a Smart Future

Increasingly growing social needs and the abundance of city opportunities has caused a large migration from rural areas to cities and speeded up the process of urbanization since the beginning of last century. Looking at the ratio indicates that large cities are covered only 2 % of earth surface where half of the entire world population is living in cities. These urban residents are also responsible for consumption of 75 % percent of the total energy and they cause 80 % of the total carbon emissions. Development in technology increased people's ambition and expectations [10]. They are willing to travel more, spend more, and consume more. Taking availability of collected data via smart devices and city sensors into account, development of a well-designed organization that can be provided by streams of data channeled through Information and Communication Technologies (ICT) is essential.

The key concept of such organization is to supply correlation between systems and subsystems with ICT collected dataset. Developing innovative urban governance, which takes advantage of available data and its analytical outcome, is an essential requirement in order to provide both efficient and effective governance. Availability of such an urban governance structure enables governments to become smarter in sensing changes in their urban environment, consequently, create smart solutions for their emerging city problems, and more importantly, learn from their past decisions and actions.

Nowadays, cities like London, New York, Hong Kong, Singapore, and many other developed cities are well aware of the key role ICT plays in urban governance and they are using their respective organizations effectively. ICT helps all stakeholders of the system and its users at large to reach instant and accurate information in real-time and continuously. It also improves the correlation among other systems. The correlation of subsystems not only improves the efficiency of the systems but also improves the quality of life for citizens at large in dealing with urban issues including but not limited to the ones discussed in following.

As a result of uncontrolled urbanization, limited sources of the world are decreasing with an increasing rate and that is affecting quality life concurrently. A United Kingdom based research indicates, 25 % of people who are living in urban neighborhoods believe that their area is getting worse opposing the opinion of only 0.1 % of them who believe that it is getting better [11]. Assuming that is a trend in similar urban structures, big cities do not necessarily satisfy their citizens when it comes to required standards of urban life.

Education is another important factor for creating sustainable future. People who migrate from rural areas to cities often have bigger families. In rural areas, number of people in a family is traditionally considered an extra power as it is translated to workforce, which is evidently the opposite in case of cities. People who live in cities want to have less number of kids so that they can provide them better opportunities for education and growth to ensure quality of life in the long term. The average population size of standard classes at schools for example is 15–20 kids in a city [12]. However, rapid immigration force schools to accept more students that will cause a new wave of urban issues.

At the beginning, building new schools can be seen as a solution while finding experienced teachers for serving those schools is not always possible or is at times very challenging. This solution however, in long term, will result with a decrease in education quality. To address such issues more effectively, smart technologies need to bring into educational methodologies before it is too late. For instance, replacement of standard boards with smart or virtual boards, usage of visual representations, and Internet can help us to provide more realistic systems for education and meet the expectations beyond population size limitations.

Conditions of health system can also be affected negatively in the wake of rapid urbanization. In many countries, particularly at USA, health system is among the most complex urban systems and extremely hard to be managed. In large cities, especially in developing countries, long lines in the hospitals cause problems and doctors cannot get enough time to take care of their patients. This may cause wrong diagnosis and maltreatment. As a result, long-term failure in urban life becomes inevitable, as unhealthy people are costly for the society on top of the fact that they may not be able to work and contribute to the system financially. From a different perspective, a health system suffering from certain ineffectiveness can cause a national economic failure.

Transportation is another major urban issue. Traffic density is directly proportional with population in a city. Many big cities put limits on the time of using personal cars. In some highly populated cities such as London and New York, pedestrian culture is growing. The most important precaution in general, is to encourage people to choose public transportation. Innovative changes can improve service level of public transportation as an incentive for using them more. According to The World Bank, China is the most populous country in the world with over 1.3 billion people. Efficient use of public transportation, hence, is a common adoptive policy. In some other countries like USA, the culture of using public transportation is yet to become prevalent. Institutionalization of such culture will save a lot of time for contributing citizens, supposedly through elevating traffic jams, saving fuel consumption, and many more factors, which will have a decreasing impact on pollution.

Availability of smart and efficient transportation systems is definitely an incentive that might encourage people to choose them over using personal automobiles. It can save energy and money by decreasing collective travel time. Instant and accurate data sharing as one aspect of a smart solution can increase the integration between different travel options. This will help people to spend their time more productively and time incentives might be a huge encouragement for using public transportation in long term.

Moreover, increase in population could possibly cause other issues such as security problems, environmental pollution, energy crisis, which all threaten the sustainability of the society. At a larger scale, it can cause environmental disasters all around the world. Environment has no boundaries. Therefore, any disaster, which happens in any part of the world, could possibly affect other countries. As an example, the nuclear accident in Japan in 2011, caused radiation in California across the Pacific Ocean [13]. This is why at a higher level, integration among systems at a transnational level is also very important. This will make all countries in the world responsible and preferably accountable regarding environmental issues.

5 Smart Solutions for Smart Cities

Smart cities are defined in the literature as "the use of smart computing technologies to make the critical infrastructure components and services of a city—which include city administration, education, healthcare, public safety, real estate, transportation, and utilities—more intelligent, interconnected, and efficient [14]." From another perspective, smart cities are economically developed environments inclusive of their residents, who are in constant interactions with them. The usage of ICT and connectivity of all subsystems is considered among major concepts in the context of smart cities. Smart systems are by nature, complex systems. In complex systems, subsystems are tightly coupled and any error in one section may destroy the whole [15]. Complex systems are dynamic as their inputs and outputs can change rapidly according to stream of data. On a positive note, however, storage of previous data in such complex systems can create memory for the system they evolve to at future states.

Subsystems of a complex environment have non-linear relations and recovery of one component may create a positive or negative effect on others. New solutions in such environments should be established on stored data from previous states of the system. Smart cities consist of different subsystems coordinated with each other. To conduct research on the understudy case, we categorized these subsystems under seven different sections as (1) governance, (2) economics, (3) smart living, (4) energy, (5) communication, (6) waste and (7) culture. All of these systems need to be integrated and governed effectively. Showing these systems in seven different groups neither assumes independence of these systems nor claims exclusiveness to these categories. The categories have only been selected for the purpose of research organization [9].

According to such categorization, city governance is chosen to be the main focus and owner of all other systems. Governance in fact serves all other categories and is directly affected by their outputs. The main objective of governance in a smart environment is to create integration among all systems and assure accuracy of information flow. New projects need to be designed, new systems are required to be integrated to current ones and latest technology needs to be well understood to meet the necessities of the environment. Such directing guidelines enable stakeholders to use and relay on the entire system and increase their participation. Figure 1 illustrates an arbitrary list of stakeholders within smart cities, identified for the purpose of the case study.

Governments are supposed to encourage stakeholders to be a participant and contributing part of their smart city. For example, consider a roadwork on a certain urban area. If an integrated smart network were already in use, an individual would be able to see this from her smart device (phone, tablet, laptop, etc.) using a particular applications developed for the city or general ones and chooses her direction and mode of transportation based on that information. She would be able to use the same venue to report a problem as she goes on the road. Her smart device would communicate the geographical coordination to the governing system directly and the importance of the issue could be evaluated by the system automatically based on the number of similar reports using the same or similar applications.

In the case of urban emergencies such as fire or traffic accident too, individuals could be able to inform the governing body as well as other citizens about the situation, using the same smart devices. The central smart system would possibly collect all these reports and send alert to other citizens who would use urban applications on their smart devices or through public broadcasting system to those

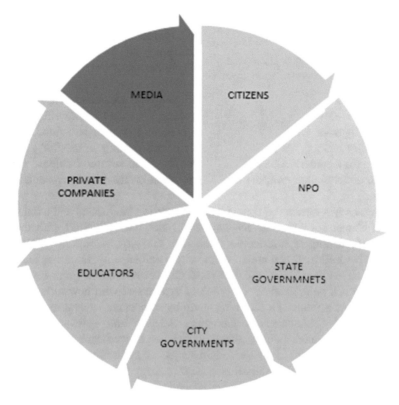

Fig. 1 Stakeholders of a smart city

who might be listening to radio or watching television programs. The significant point is to verify the accuracy of information, which could be done in variety of ways including the crowd-sourcing test, i.e. the frequency of receiving the same message from other citizens at the same location or on the same topic. After assuring accuracy the necessary actions will be taken and the appropriate messages will be sent. Sustaining a supply of instant information flow will provide us with some level of intelligent governance.

Obviously, there might be lots of other ways through which "smart" governance can be realized and applied. Automatic alerting system is only an example that along with lots of other concepts such as openness, transparency, public access, reliability, e-governance, data tracking and record keeping, etc. creates the structure of innovative governance for smart cities. In parallel to the concept of governing smart we have the challenge of living smart. Again, one might categorize such idea in many different ways. For simplification purposes the major categories of smart living are summarized in the following eight major sections: (1) education, (2) transportation, (3) healthcare, (4) safety and security, (5) energy, (6) communication, (7) waste, and (8) culture. Each of these categories is discussed in more details in the following.

Education is at the heart of any learning process from kindergarten throughout university. Citizens with higher level of education are more capable of using smart solutions and smart toolset offered by their city and this might hopefully provided them with opportunities to make more intelligent decisions and live a smart life. In a more particular level, higher educational institutes can par take in educating citizens on the concepts of smart cities and making data driven decisions using applications on their smart devices. Such smart city projects are usually more effective when conducted together with universities. The case of Hoboken is a good example. Some faculty members at SSE along with students residing at the SIT campus in Hoboken, from undergraduate and graduate levels at several faculties, started a project in collaboration with the local government on the way to create a smarter Hoboken.

The researches were to be conducted at immersion lab located and administered by "The Center for Complex Systems and Enterprises (CCSE)" with the main purpose of creating a sustainable future for the city. According to Hoboken Demographic Information reported by the city's municipality, the urban area under study, which includes about 50,000 residents, could make a great test bed for theories and can be applied as a city lab to experiment with new models of governance. SIT as a result of a far-going study by students of several research teams from Engineering Management, Computer Science, and Software Engineering departments, generated a smart phone application, named Smart City: Hoboken. This is an ongoing research and the main subject of this study is to cover different urban systems such as: transportation systems, energy systems, telecommunication systems, water and sewage systems, emergency services, postal services and analyze interactions among them.

Smart City: Hoboken is a result of synergy between different stakeholders, such as NJ Transit, Port Authority of New York/New Jersey (NY/NJ), utility organizations of New Jersey (PSEG), local governments and members of SIT. These entities are cooperating to make Hoboken a smarter city. The aim of the project is to provide information to citizens or application users, who are most probably work in the city but reside elsewhere, by channeling processed information collected by city sensors and smart meters as well as crowd-sourcing information collected by citizens. People are supposed to reach this information from a website or directly connect to the application installed on their smart devices, ultimately. To make such vision to reality, hundreds of sensors are being installed.

With this technology, traffic information, energy consumption, telecommunication, emergency services, infrastructures, safety and environmental sustainability can be measured and can be monitored. At this point, early academic studies are important to examine past data, to have a good knowledge about latest technology and to generate new ideas, which are appropriate for Hoboken's lifestyle. SIT research team tries to understand the behavior of citizens and how they can adapt to or use technology to reach project goals. It is important to understand the future of smart cities and using real cases such as Hoboken is crucial in this endeavor. We can put whatever technology we want in cities, however, it is not clear that citizens are benefiting from the outcomes of the system or the extent to which they might benefit. It is crucial to figure out if this is all going to be luxury or a necessity of life.

Transportation is the center of attention in many urban studies. As a result of increased population, it has become very difficult to keep a car in cities. Especially in highly populated cities like New York City, London, Mexico City, Bogota, Tokyo, Tehran, and Istanbul, it could take hours to move 10 miles road in rush hours. Both vehicles and pedestrians spend a lot of time trying to reach from one point to the other. In order to decrease high traffic density people need to be motivated to use public transportation. This requires the installment of an effective incentive structure and a plan to promote the social engagement.

To encourage public transportation, public vehicles need to be well designed and be comfortable. Another incentive could be availability. The schedules of public vehicles need to be well arranged and effective. Some kind of penalty could also be considered for those who refuse using public transportation. "Congestion Charge" in London is an example of such penalties. If an individual would like to use personal car in Central London between the hours 7:00 am to 18:00 pm, they need to pay £11.50. As a result of applying this fee, waiting times for bus service fell 30 % and a 17 % drop in total traffic was observed. According to Seattle Transit Plan Briefing Book, less traffic means less consumption of fossil fuel and consequently less CO_2 emissions.

Better traffic management can save lots of time. This collective save of time can be used toward productivity in the society. Moreover, decreasing traffic levels may save lives. Firstly, less traffic means less likelihood of accidents. Secondly, all around the world, lots of people lose their lives because ambulances or other emergency vehicles cannot attend to them quickly enough due to traffic jams. A well-managed transportation system facilitates performance of other urban systems as well.

Healthcare and health systems as a major urban subsystem are subject to vulnerabilities for variety of reasons. Regardless of urban levels of development, growing population increases risks of facing health issues. Productions of new and advanced drugs are precluded by increased costs in production. Besides, there is more likely to have a loosened network in health sector. Healthcare stakeholders such as doctors, patients, and insurance companies can share their flow of information actively and effectively only when there is developed a strong integration among all subsystems and components. A seamless integration will save time and contributes to decrease of total cost in collective healthcare system. Smart technologies will need to meet the increasing demand in this realm.

Moreover, there should be an educational program that informs stakeholders of healthcare system as well as citizens at large to respond timely to the increasing demand. To enable the required level of integration, smart healthcare systems are using cloud-computing features as data storage to facilitate access to all data by as well as other health providers. It increases the system transparency and increases the level of the health system by sharing cases between institutes. Smart cards are examples of such systems through which health records of patients are uploaded on cloud and become visible to others who have permit to access. Every time a patient visits a health unit her information will get updated. Patients can use the same system to pay their hospital bills as well. During an emergency, these cards can save lives with immediate information sharing capabilities.

Safety and Security are inseparable concepts of urban life as are of technology. Reaching any type of sustainability is an impossible task where safety and security do not exist. Ensuring the security of a growing population while keeping the system under control requires an innovative approach. Safety of a city is valuable when achieved proactively as opposed to being limited to catching criminals. The most effective way for keeping security proactively is when data is collected and processed in real-time and is being disseminated through reliable channels and continuously. The process of collection and analysis can be done locally to assure accuracy and efficiency of response. This data should be accessible for relevant departments such as police and firefighters, etc. According to IBM, New York City for example was successful in decreasing crime rate by 27 % just by speeding up the processes using centered real-time data.

Energy infrastructure is the backbone of interconnectedness among all urban systems. In the case of smart cities, energy grid, source of energy, rate of consumption, peak times, and etc. are required to be analyzed to achieve efficiency and effectiveness in usage. All other parts of urban systems including transportation, healthcare, communication, governance, etc. cannot work properly without a reliable energy system. That is why to achieve a satisfactory level of optimization, considering energy behavior of citizens should be the focal point of attention in innovative governing methods of future cities.

Communication is another crucial subsystem of urban environment without which any connection among the other components becomes very difficult. For any smart system, the accuracy of information is the key and without having a reliable communication it is very difficult to keep the environment at a required level of awareness. The Internet is becoming the favorite mode of communication in the recent decades. E-mail, social media services, such as Twitter, Facebook and LinkedIn started to take over the modes of communication and substituting phone calls. To supply rapid Internet, cities need to have high quality infrastructure for 2G, 3G, and Wi-Fi services.

City governance can communicate and inform citizens using the same tools including but not limited to online social networks, text messages or e-mail alerts. For example, Hoboken has a smart phone application called, "HOBOKEN311[1]" through which communicates with residents using a one-way channel. This application is designed to communicate non-emergency issues with citizens. This enables citizens and visitors to reach out information about parking, transportation and city news. Users can also sign up for receiving frequent messages regarding city matters. Besides these smart applications or email channels, traditional media

---

[1]http://www.hobokennj.org/311/.

such as television, radio channels or local newspapers can also be used for similar purposes as well as advertising city attractions.

Waste and the more general topic of managing urban waste is becoming one of the most important factors of intelligent governance. As reported in an article for Harvard Business School "one man's trash is another man's treasure [16]." The same could be applied to similar concepts within the context of cities. If municipals govern their waste efficiently, they can make profit through financial transactions in waste market or transforming waste into energy. Collection, processing, energy recovery and disposal processes of waste can lead a city to become smarter. Countries like Sweden, Austria, and Turkey already have established facilities to create bioenergy from trash.

According to a piece on NY Daily News, Sweden leads other countries on recycling, taking advantage of solar energy along other sustainable projects. Sweden obtains energy from garbage for electricity and heating and they have already run out of garbage since 2013. This forced Sweden to import garbage from Norway [17]. This is a good example on how recycling could be an important part of sustainability plans. At this point, the key to success is to keep citizens informed on issues of recycling and sustainability.

Culture and cultural aspects are also essential for increasing the rate of success in application of smart solutions as well as in adoption of new paradigms of governance. There is no universal method that could be used for all cases. Each city should develop its own customized project taking all of local circumstances and limitations into account. The most effective governing plan is the one that is tailored according cultural considerations of the jurisdiction upon which the governance is going to be applied. Local governments are responsible for understanding the requirements of citizens and creating effective and innovative solutions to maximize city benefits for all stakeholders of the city.

6 Incentivizing Sustainable Behavior

According to data gathered and reported by International Energy Agency (World Energy Outlook Fact Sheet in Paris, France[2]) energy demand will increase by one-third from 2011 to 2035. As a part of legacy system however, the old-fashioned energy grids are still widely in use. It becomes more apparent that traditional systems for distribution of energy through electrical grids are not sufficient for supplying demand of the 21st century. One solution to this legacy system issue is to increase usage of smart technologies including renewables smart meters and smart grids. Another approach is to influence collective patterns of human behavior to encourage sustainable trends in all sections of society.

---

[2]http://www.iea.org/media/files/WEO2013_factsheets.pdf.

Human behaviors are systemic outcome of a multivariate function of: social structures, institutional contexts as well as cultural norms, among others. Socio-structural and socio-technical networks along with individuals, who are the using agents of such structures, can equally impact adoption or change of collective behavior. In practice, social structures form rules and resources to organize, guide, and regulate individual actions while, individual actions can create and change social systems [18]. For instance, household environmental attitudes play an essential role in energy consumption behavior. In other words, behavior is affected by variables of individual attitudes and contextual variables such as interpersonal influences, regulations, interventions, institutional factors, incentives, constraints, knowledge and skills. Other aspects such as moral norms, information, and communication can significantly impact energy behavior [19].

The relationships between individual and cultural aspects affect robustness and resilience of behaviors [20, 21]. In the case of energy behavior, moral utility that identifies collective beliefs, values, attitudes, routines, norms, self-efficacy, constraints, and habits play a key role in incentivizing sustainable behavior. On the other hand, physical and structural conditions including home size, room size, widow area, technology, and standards positively influence information on energy consumption behavior and residential households' demands for energy. A socio-demographic characteristic including households' income, age-compositions, and education level affects the residential energy consumption [22]. Financial incentives are important too. Rewarding would encourage consumers to positively change their energy behaviors. Economics utilizes rational choice theory to analyze patterns of collective behavior. According to the theory of choice, self-concern is a basis of individuals' decisions. Sociologists focus on the relationships from a different perspective in which there is a bidirectional relation between social context and individuals' choice.

Social context plays a key role in defining the individuals' needs, attitudes, and expectations about social norms, technologies, infrastructures and institutions. Sociologists believe that energy consumption and provision of energy resources do not determine individuals' decisions. It is stated that, "the difficulty for individuals to change their consumption patterns is also highlighted since lifestyle and use of material goods construct meanings and identities, which account for individual social expectations (social norms), self-expectation (positive or negative outcome of saving energy) and self-efficacy (perceived effort's effect to save energy)" [23]. Both social norms and self-efficacy positively impact the reduction of energy consumption. Socio-technical forces affect individuals' habits, which in turn influence energy consumptions practices. Moreover, limitation of awareness about economic incentives (e.g., subsidies), limitation of capital incentives for energy-efficient equipment and limitation of knowledge about regulatory policies impact the individuals' energy behavior change [23].

Technology, as discussed above, is another important factor for behavioral change and consequently decreases of energy consumption. For instance, smart meters can be used as an energy-monitoring tool to increase consumers' engagement and information on energy consumption. By utilizing energy meters householders have more control on their energy consumptions and costs [24]. For example, capital-intensive petrochemical companies changes the fossil-fuel-based energy system to non-hydrocarbon energy technologies since fossil fuels are depleted and the costs of extracting fossil fuels are becoming too high. Under the right circumstances, public authorities are able to transit smoothly to new environmental technologies to achieve sustainable energy in future [25].

7   A Game Oriented Approach

In this section, a game oriented approach is presented regarding influencing energy behavior of Hoboken citizens. The research was conducted in the form of an academic project for SIT campus residents in Hoboken. The energy consumption, particularly on usage of electricity was decided to be the core of the research as it was easier and more relatable from the students' perspective. Moreover, the energy consumption in general and usage of electricity in particular are the core variables of consumption patterns and indicative of such behavioral trends in regards to sustainability studies. With that consideration, in order to create the right contexts for this research, it is necessary to have an analysis on usage of renewable energy around the world. Renewable sources are those not containing coal, natural gas, petroleum and other fossil fuels. These can renew themselves without needing any human interference. In order to achieve sustainability, countries need to invest on renewable resources. Results of this research are listed in Figs. 2, 3 and 4.

Considering difficulties of using renewable energies, it is crucial to adopt hybrid strategies. Smart meters are integral parts of many intelligent systems. Energy consumption data becomes visible to the consumers with help of smart meters. According to the data from General Electric, using smart meters in 10 % of houses

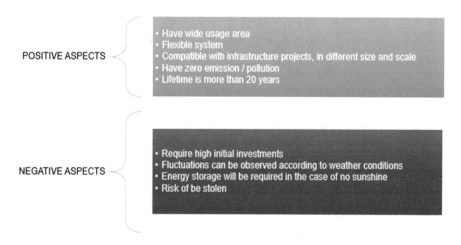

Fig. 2 Solar panels

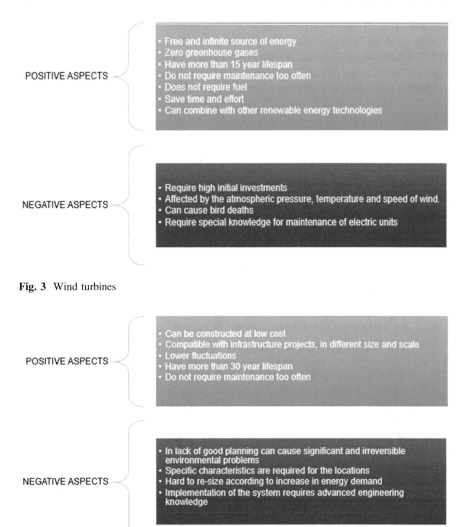

Fig. 3 Wind turbines

Fig. 4 Hydroelectric power

in the USA can prevent 3.5 million tons of carbon emissions. Increasing the number of smart meters will continue to reduce the amount of emissions. Smart grid application is also a new technology but spread lots of countries. Smart grids allow communication between manufacturers, suppliers and consumers. They also, enables consumers to see both real-time consumption data and rates of hourly electricity prices with help of smart meters and they can adjust their consumption or its timing, consequently. Smart devices can communicate internally and lead smart buildings, which are controlled, from one system. With these technologies, dependence on fossil fuels can be reduced (Fig. 5).

Fig. 5 Transmission of technology

Smart devices and newly developed applications are also very important in the process of behavioral change, as they have become an inseparable part of daily life, in addition to a source of entertainment. This coupled with the natural inclination of people for playing games and engage in competitions makes these devices a great tool for incentivizing sustainable behavior. Probably that is why game-based competitions using smart devices applications are increasingly in use. For instance, "Saving the World" that is in draft, is focused on competition to create awareness. A similar approach should be adopted for change of behavior through three stages: transfer of consumption data into game, administering the game, and disseminating energy saving tips. An example of this approach is used in the following for the case of Hoboken.

8 Using Data in the Game

8.1 Transfer of Data into Game

At the inceptive stages before receiving continuous streaming of data from smart meters, some initial data is required to be transferred into the game manually. At the end of each month, invoices information and the total amount of electricity consumed (in kWh) should be considered to determine the score in the game without exposing any private information from the users. However, the player can leave the location button on and help the programmer to gather information on consumers' behavior in different locations. After each player registers to the game, she is given the opportunity to select an avatar for herself. In addition, the player needs to create a home design, which has similar features of her apartment in real life. At this stage, it is important that the player add all electric devices that she has or use in real life and also enter the number of people living at her home (or her residence, which could be a room they share at a dormitory on campus) into the game. Then, the daily life of the player begins. For example, if a player put water into a kettle and makes tea, she should add using her avatar choosing how much water she put into the kettle, etc. Likewise, all consumption information can be transferred to the game manually.

9 Playing the Game

It is essential for the avatar, as she plays, to do the same activities of the player in her real life setup as much as possible. The game is designed in a way, which for every piece of information about consumption, the player will accordingly gain more points. For example, when the player goes to work and turns on his computer she can choose plugged or unplugged options. The player can change this option during the game. The information on consumption will be multiplied with points and total score will determine daily score and ranking. Determination of monthly score is highly related on how consciously the player chooses to behave. In reality, each activity has an optimum range. If a player fills a big kettle just for one cup of tea or goes to a very close place by car instead of walking or cycling, she is moving out from the game's optimum values. Moreover, the monthly invoice is the biggest proof of her behavioral choices. If the amount on the bill is very different than the game values, the player will lose points. If the player chooses to share her location information, she can see her ranking, in comparison to other residents of her street as well as the city baseline.

10 Disseminating Energy Saving Tips

Energy saving is the key in decreasing the total amount of usage just through changing behavioral patterns without affecting the production areas, economic development and quality of life. In other words, energy saving is preventing the waste of gas, electricity and water, obtaining recycling of solid waste also taking advantage of using renewable and alternative sources by using more efficient sources and decreasing unnecessary energy demand. Comparable to industrial sectors, households also have a critical role in the consumption of energy. Energy is essential for all life activities, however controlling percentage and timing of usage could be possible through affecting behavioral patterns of individuals. Figure 6

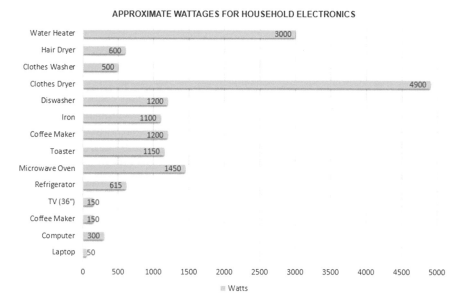

Fig. 6 Consumption profiles of households electronics

depicts the energy consumption data of some household electronics as an example. Unconscious consumption of energy can cause energy shortages and increases in energy prices, which is harmful for both the balance of natural resources and the household economy. That is where personal choices become important both in choosing behavior or picking energy efficient devices.

11 Goals of the Application

The main purpose of the proposed application is to provide a platform where people can have fun and at the same time, learn about energy consumption concerns, just as one of the costly variables of energy behavior. Knowing that we are dealing with a large-scale system in a city, if little changes can contribute to a little percentage of saving per household, the collective saving will be huge for the entire society. Such amazing outcome can be achieved through the following phases. Firstly, real-time information is possible when suing the application as opposed to regular monthly bills. The continuity of receiving information will help consumers similarly to how it does in case of a diet program. When they receive a notice over consumption, they will still have the opportunity to balance it out in the following days. Instantaneous information about consumption data can be visible to the consumers with the help of sensors, smart meters and monitoring technologies through their smart devices using customized dashboards. Daily energy assessment will be not complete with energy savings tips and consumption data in kWh. But also, the game will supply information about how much trees we save, how much less CO_2 we generate and how we prevent global warming as well. In this way, targets of this application will be more understandable and can reach more people from different cultures or education levels.

Secondly, having partnership with other big companies can help this game to become more popular. Phone service providers such as Apple or Samsung can help players to collect points and redeem them at online stores for applications of their choice. Thirdly, information collected on other energy sources such as gas and water as well as transportation could be added to the application, giving players information on CO_2 emission they cause with their individual choices. Finally, the ultimate goal of using an application is to eliminate the need for manual data entries and to create a real-time streaming of data to the players. Moreover, the collection of all data through this application is supposed to be used for academic research.

12 The Case of Hoboken

In this section, a survey-based research conducted in the city of Hoboken, particularly for the resident students on campus of SIT, is presented. Energy consumption in the residential buildings is directly proportional with its residents' behavior. The residents' choices on electronic devises and consumption habits affect the amount of consumption. As a part of this study, students who live in residential buildings of SIT campus were observed for months. The campus is located in Hoboken, a city on the borderline of two states of NY and NJ at the East Coast of USA. Hoboken's location along with lots of other characteristics of this little city has made it a good candidate to become a laboratory for testing smart solutions. On the way of becoming a smart city, we need to have smart residents and that has been the reason behind choosing this location for our experiment.

SIT campus buildings are mostly located within the boundaries of the university. However, some of the residential buildings of the institute are off-campus. In fact, most of the building we picked for our experiment is off-campus but still within the boundaries of Hoboken. Being a part of the city, these buildings have better understanding of life style in Hoboken. Off-campus buildings are located on regular apartments in the city and are connected to the building management administrations, which do not belong to SIT. There are many different off-campus buildings both for undergraduate and graduate students. Eight of these buildings located at: 538 Washington Street; 800 Madison Street; The Curling Club; 828 Garden Street; The Shipyard; 733 Jefferson Street; Juliana Apartments, and Patrician Apartments were analyzed within the scope of this research. Typically, four people share an apartment and apartments consist of two bedrooms, two bathrooms, a kitchen and a living room. In addition, dishwasher, washing machine, dryer, heating and cooling units are available for each of these apartments. Residence Life of Stevens Institute of Technology designed different projects to encourage students consume less energy and try to turn this to a competition between different buildings as well.

During this observation several data was collected from two different sources to check validity. First dataset came from PSEG, the utility company that serves NJ exclusively providing the state with water, electricity and gas. The bills issued by PSEG for different buildings and apartments were used as one set of data. The second dataset was gathered directly from residents of these buildings through a survey.[3] These eight buildings, which were the subject of this research, have different number of apartments for SIT students. Therefore, the average values for each dataset were considered. In this experiment data from a total of 122 apartments in these eight buildings were collected. In addition, electric bills were analyzed from the months of July of 2013 to February of 2014. July represented, as 1st month and consequently February was the 8th. During these 8 months, there was a chance to observe electric bills for different conditions from the weather's perspective. For each of these apartments, bills were collected and the average amounts per month were calculated throughout the entire period of the eight months. The results for "average bill per building" are shown in the Fig. 7.

The review of results shows that Shipyard has the highest consumption amount whereas Juliana is the winner of this particular game with less consumption. At this point, there is a $131.72 difference between the average values of these two buildings per month. This is because Juliana is a 5-year-old building where Shipyard is almost 15 years old. Shipyard is one of the oldest buildings in residence halls, and thus equipped by old technology. Air Conditioner (AC) units in Shipyard

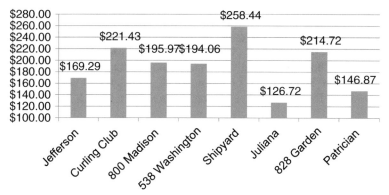

Fig. 7 Average cost of consumption

---

[3]The Residence Hall Survey is presented at the end of this chapter as an appendix.

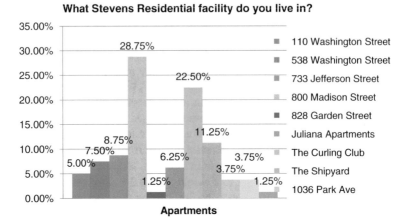

Fig. 8 Residential locations of students

are very old and even do not have the option of automatic setting. As a result, most of the times, it is either too hot or too cold in rooms and room temperature cannot stabilize. This is highly inefficient and resulted with high consumption as a result.

Taking a look at the survey results is also interesting in this case. Students, who live in off-campus buildings of SIT, filled out the survey. A total of 80 responses were received and analyzed in the scope of this research. First question specifies the building the resident belongs to and the results of the rest of questions are presented in Fig. 8. Second question was related with the number of occupants in an apartment, the answer to which shows, 70 % of these students live in a group of four people together while 23.75 % of students live in a group of three. This almost follows the same pattern with Hoboken citizens, who are consisting of one couple with one or two children in average.

The 3rd question was asking about the light bulbs that students use in their apartments. Most of the students, not surprisingly, just using what were supplied by the building without being aware of energy requirement of them. Most of these apartments have hybrid types of bulbs which includes traditional incandescent bulbs, compact fluorescent (CFL), spot halogen bulb and Light-Emitting-Diode (LED).

Halogen bulbs are very inefficient and produce much of the electro-magnetic energy in the infrared spectrum. This causes them to emit large amounts of heat as opposed to visible light. They are hot to the touch as opposed to LED or compact fluorescent lighting, which produce primarily visible light. It is not unlikely to see over 1000 W of halogen lights in a kitchen providing the same quantity of lighting in Candela as 150 W of incandescent or 30 W of LED lighting. To achieve a better result in terms of consumption, CFL bulbs can replace much of the traditional bulbs. CFL bulbs come in a variety of sizes, power ratings and colors, which can closely replicate the effect of traditional bulbs while supplying superb saving benefits. The downside from the data is that LED lighting has not taken off yet.

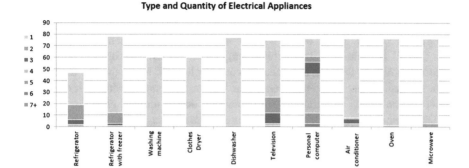

Fig. 9 Types and quality of electrical device

LED lighting is far more efficient than even CFL but unfortunately the significant down costs of LED lighting make it a risky option for many apartments.

Next question is related about electrical appliances, which students use in their apartments. By looking at the Fig. 9, we begin to understand reasons behind extensive power consumption. Nearly all respondents reported to having a refrigerator/freezer. Other possible causes of excess consumption are laundry machines, dryers, dishwashers, air conditioner, electrical ovens and microwaves. The interesting dynamic is that while some appliances only appear once or twice per apartment (such as washing machines, dryers, dishwashers and ovens) other appliances appeared in large numbers. These include, refrigerators, freezers, televisions, computers and microwaves. When both lists are combined, we see that there is appliances both appear in large quantities and consume a lot of power. These are refrigerators, freezers, air conditioners and microwaves. Based on how much time these devices are being used, they could potentially pose a great deal of problems in terms of consumption. They are both very common in the apartments in consume a lot of power.

Ironically, a separate class of appliances, namely computers and television sets are problematic, despite the fact that they do not consume a lot of energy, only because they are very common and often on for excessively long periods of time in all apartments. Television sets and computers are mostly used for long hours. These devices may seem harmless but their cumulative effect over the entire day is as significant as the big consumption of refrigerators, dishwashers, dryers, etc. The worst of all in the list of devices are those that not only common in the apartments and consume a lot of energy, but are also used for extended periods of time. The primary device in this class is AC. The air conditioning appliances can single handedly impact electricity bills during hot season.

It was difficult to obtain specific information through 5th question by asks: "What is the brand and electric consumption of the electronic goods?" This could partially because it must have been difficult to read labeling on some of old gadgets used in the buildings under study. For refrigerators and freezers, power consumption correlates with the size of the units. Small units can consume 250 W or less. Larger units may require over a kilowatt of power. Dryers for clothing proved to be a significant consumer. One unit was reported at 500 W while another was reported at 3.5 kW. Based on the power consumption of modern units, it is unlikely that 500 W is representative of the average. The 3.5 kW is more likely to be an accurate number. Despite the specific consumption, it is acceptable to say that dryers consume an incredible amount of power and can drastically increase the power consumption of an apartment depending on the number of loads.

Dishwashers also proved to be high consumers of energy. The data shows that a consumption rate of 500 W to 1.5 kW is expected from a dishwasher. Once again, depending on the number of loads and duration usage per load, dishwashers can represent a significant increase to the energy consumption of an apartment. The same is true about electrical ovens (>1 kW), microwaves (1 kW) and air conditioners (<800 W).

The data also identifies the lower power devices such as television set and computer. Television consumption is typically limited to fewer than 100 W so the problems, as discussed previously, comes from the amount of time these devices are used. For example, 10 h of leaving a television set on is equivalent to roughly 1 h of using a dishwasher. Computers ranked second in the list of low power devices. Many laptops consume less than 50 W with more powerful units consuming close to 100 W but rarely over 150 W. Desktops are exceptions. Based on the responses from students, it is clear that many with desktops consume several hundred watts of power. Occasionally, a desktop with 1000 W or more can be seen in the list of appliances.

The 6th and 7th questions are about sources of energy for cooking and hours of usage. Data proves that despite large power consumption by electric cooking devices, such kitchen appliances do not represent a large portion of the electrical energy consumed at the apartments. This is most likely because most respondents reported using the stoves no more than 30 min. This represents approximately half a kW per day. Just one person achieves the same consumption rate after 5 h per day of laptop usage. Despite the large power requirement, electric stoves are not used long enough to represent a significant portion of overall electric energy demand for our population.

Question 8 is asking about the number of hours students turn on their air conditioning appliances (AC) per day both for cooling and heating purposes. Figure 10 represents the results based on which most of respondents reported using their AC systems between 12 and 24 h a day. This is a significant portion of electric consumption because these devices consume hundreds of watts and are often being used for hours per day. While the average oven was responsible for half a kW per day, an AC unit might be responsible for 10 kW or more over the course of a 24-hour period. This data from the participating students in the survey, quantifies a well-established correlation of excessive AC usage with large energy consumption. Despite representing a lower portion of electrical consumption than AC units, lighting plays a significant role because of the number of hours that light bulbs are being kept on.

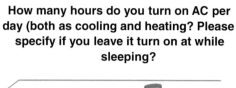

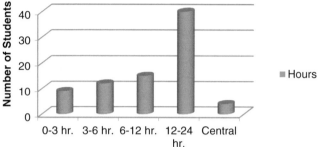

Fig. 10 The number of hours air conditioning systems usage

According to collected answers to 9th and 10th questions, nearly two thirds of students reported usage of lighting over 6 h per day. Over a quarter kept the lights on for more than half a day and 10 % admitted to regularly leaving the lights on at all times. Even though lighting does not consume a lot of power, when used over long periods of time, large quantities of energy are consumed. This is why efficient lighting is crucial. The lights can be used for long periods of time and consume very little energy if LED or CFL light bulbs were used.

As for 11th question, collected answers indicate that the average temperature for a room in both winter and summer was kept around 70 °F. The temperature setting on cooling and heating devices is important indicator of power consumption. Lower temperature settings in summer and higher settings in winter results in increased consumption. According to SIT Residence Halls, it is suggested that temperature settings be kept around 75° for summer and 65° for winter seasons. Overwhelmingly, 95 % of students reported maintaining sealed doors and windows during cooling and heating. Based on this rather conclusive data, we can assume that energy leaks to the outside are not a primary cause of increased energy consumption. Over 80 % of the students reported usage both dishwashers and washing machines at full capacity. This shows that students usually turn these devices on when needed.

The 15th question is asking whether students switch off their screens of computer, television, game console, or similar devices after using them. According to collected answers, 75 % of the students reported that they turn their screens off after using them. Screens that are left on for extended periods of time consume a significant amount of energy. Still, nearly a quarter of students reportedly do not pay attention in this regard. Last two questions were asking about availability of students to participate in future studies.

The statistics shows 66 % of the students are willing to participate in energy saving activities. However, there are still a rather large percentage of respondents showed no interest in such activities. This shows the necessity of an educational plan for increasing levels of engagement among citizens. Developing a monitoring program might be used for this purpose as it brings awareness to the matter and also creates competition among students. In designing incentive plans, creating a balance between reward and punishment is highly important. It is hard to choose a winning strategy that can certainly dominate the others. Results of the survey show that 77.9 % of students believe that if they were able to monitor their energy consumption, it would encourage them to save energy.

For educated users in general, a well-designed project should consists of three phases in the technical sense: (1) installation of smart meters to the buildings, (2) transferring information taken from meters to an analysis center, and (3) presenting the results and analytics via a website or/and application.

13   Conclusion and Future Work

In the smart world of smart cities, all components of the urban system need to be smart and capable of learning. This includes stakeholders, system users, and governance structure as well as private and nonprofit sectors. Collaboration among stakeholders will increase the system reliability and effectiveness. Such connectivity is fortunately possible through the advancements and availability of technology nowadays. Collaborative actions among urban stakeholders are also known to lead up to a sustainable development throughout the environment. It requires a detailed plan for influencing sustainable behavior in city networks among many other policies set and enforce by urban governing bodies in charge. This is where innovative urban governance can effectively influence desired impact through technological tools such as city sensors and smart devices along with governing methodic approaches through policymaking, and leveraging urban rules and regulations.

One of these effective approaches as identified by the literature is development of games and utilizes them in governing mechanism. Due to the attraction that human beings natural have toward the concept of competition, they usually enjoy participating in games. The participation often gets more heated when it is supported by an incentive structure. People usually like to score more and it is more effective when combined with challenges and competition in general. Game based approaches have been used in other fields to influence people's behavior, for example to change eating behavior of those who are interested in losing weight through a diet programs as well as in educational systems to encourage people to learn more on their own and alongside their peers. The idea here is to use the same approach to influencing collective behavior in a society toward a desired outcome. The premise for using this approach is that people will choose a sustainable alternative when given the right information and provided with incentives to do so. The same statement is indeed the hypothesis of the proposed research in this chapter.

In order to show the impact of information in behavioral changes among individuals, an experiment was designed for the students residing at Stevens Institute of Technology's dormitories on and off campus in Hoboken, NJ. The experiment was developed based on a game and the verification phase was done through a survey based data collection. The results of the experiment then was studied and evaluated. Students were supposed to respond to a survey regarding their consumption behavior. As a part of the conducted research also the results of the experiment were analyzed against similar information collected from citizens of Hoboken. The surveys consist of 17 questions regarding electronic devices students used and some other characteristics such as their speculations of the usage time.

The results of the survey clearly show that these students do not use energy in an efficient way. Despite the fact that most students are using energy with no particular consideration, survey results indicate that more than two third of them believe that monitoring instant energy consumption would lead them to save energy. It was also apparent that bringing the concepts of energy consumption to the context of an interactive game encourages the users to participate more enthusiastically. An incentive structure was developed for the game through which users could collect points. Regardless to that fact that no mechanisms were developed for redeeming the points, existence of the concept in itself was an encouragement for participation.

An exhaustive research plan has been developed to take this research to a higher level in near future. This is partially relying on the city sensors that have been installed all around Hoboken. The instruments are still under calibration and the data collection processes are under investigation. The sensing system will collect variety of data continuously throughout the city, when finalized and established. The data will then be sent to a computing center at the School of Systems and Enterprises, where they were categorized and analyzed. The outcome of such analytics will then be filtered and aggregated properly and finally disseminated among citizens of Hoboken via the application that has been developed by another team of students at Computer Science department.

The developed application can be installed on smart devices such as phones, tablets, laptops and desktops. The idea is to use this application as a channel through which citizens can communicate with the data center at SSE. The users of the application can send data of their choice and share their behavioral patterns not only in regards to energy consumption, but also on their transportation, and health related issues. The application also has the option for them to notify governing entities of the city about variety of issues such as crime, accidents, fire, pit holes on the road, and even their sentiment on a political issue.

The application will provide users with e-governance options through which they can apply for urban permits and fill out city forms. It can ultimately become a way for communication with the city government and also be used in polling and political debates. The ultimate objective of this approach, which is offering innovative governance mechanisms for urban systems, will be pursued through advancement of multi-way communications using existing technologies of sensing, data collection, analytics and visualization as well as new methodologies of crowd sourcing and civic participation. The developed governance frameworks, mechanisms, data architectures, tools, methods and processes will be presented in conferences and published in relevant journals or books.

Appendix: Residence Hall Survey

(1) Which type of Stevens Residence House you leave in? (Building name, Shipyard, Curling Club.... Etc.)

110 Washington street
538 Washington street
733 Jefferson street
800 Madison street
828 Garden street
Juliana apartments
The curling club
Other (please specify)

(2) How many occupants in your apartment?

2
3
4
5
6
Other (please specify)

(3) What kind of light bulbs do you have in your place?

Traditional incandescent bulbs
Compact fluorescent
Spot bulb
LED
Other (please specify)

(4) Which electrical appliances do you have in your place, please specify the number as well?

Refrigerator
Refrigerator with freezer
Washing machine
Drying machine
Dishwasher
Television
Personal computer

(continued)

(continued)

| Air conditioner (both cooler and heater) |
| Cooker/oven |
| Microwave |
| Other (please specify) |

(5) What is the brand (model) and electric consumption "Watt" of the item (information should be labeled behind of the item)?

| For: |
| Refrigerator |
| Refrigerator with freezer |
| Washing machine |
| Drying machine |
| Dishwasher |
| Television |
| Personal computer |
| Air conditioner (both cooler and heater) |
| Cooker/oven |
| Microwave |
| Other (please specify) |

(6) Which is the main energy source for cooking?

| Electricity |
| Gas |

(7) How many hours in a day you use oven/stove/cooker?
(8) How many hours do you turn on AC per day (both as cooler or heater)? Please specify if you leave it turn n at sleeping time?
(9) How many hours do you turn on lighting per day?
(10) Do you prefer keep lighting on when you leave the apartment or during sleeping time?

| Yes |
| No |
| Other (please specify) |

(11) What is average room temperature in your apartment? (Please just specify the degree for heater or AC)

Average temperature during the summer season (F/°C)
Average temperature during the winter season (F/°C)

(12) Are doors and windows kept closed when your heating system is on?

Yes
No

(13) Do you put on the washing machine when it's not full?

Yes
No

(14) Do you put on the dishwasher when it is not full?

Yes
No

(15) After having used a computer, television, game console, or similar, do you switch off the screen?

Sometimes
Never
Always

(16) If you would able to monitor your energy consumption, do you think it would encourage you to save energy?

Yes
No

(17) Would you be interested in participating in energy saving activities in the future?

Yes
No

References

1. Batty, M.: The New Science of Cities, p 520. The MIT Press (2013)
2. Darabi, H.R., M. Mansouri, Gorod A.: Governance of enterprise transformation: case study of the FAA NextGen project. In: 2013 8th International Conference on System of Systems Engineering (SoSE) (2013)
3. Mansouri, M., Mostashari, A.: A systemic approach to governance in extended enterprise systems. In: 2010 4th Annual Systems Conference IEEE. IEEE (2010)
4. Mostashari, A., et al.: Cognitive cities and intelligent urban governance. Netw. Ind. Q **13**(3), 4–7 (2011)
5. Darabi, H.R., Gorod, A., Mansouri, M.: Governance mechanism pillars for systems of systems. In: 2012 7th International Conference on System of Systems Engineering (SoSE) (2012)
6. Jacobs, J.: The Death and Life of Great American Cities. Random House, New York (1961)
7. Jacobs, J.: The Economy of Cities. Random House, New York (1969)
8. Barabasi, A.-L., Frangos, J.: Linked: The New Science of Networks. Perseus Books Group, New York (2002)
9. Karaca, N.I.: Smart City Hoboken: Energy Consumption Behavior of Citizens, in School of Systems and Enterprises. Stevens Institute of Technology, Hoboken, New Jersey (2015)
10. Aoun, C.: The Smart City Cornerstone: Urban Efficiency. Published by Schneider electric (2013)
11. Force, U.T., Rogers, R.G.: Towards an Urban Renaissance. Spon, London (1999)
12. Mosteller, F.: The Tennessee study of class size in the early school grades. Future Child. 113–127 (1995)
13. Takemura, T., et al.: A numerical simulation of global transport of atmospheric particles emitted from the Fukushima Daiichi Nuclear Power Plant. Sola **7**, 101–104 (2011)
14. Washburn, D., Sindhu, U.: Helping CIOs understand "smart city" initiatives. Growth. **17** (2009)
15. Buldyrev, S.V., et al.: Catastrophic cascade of failures in interdependent networks. Nature **464**(7291), 1025–1028 (2010)
16. Blanding, M.: Transforming Manufacturing Waste into Profit. Working Knowledge: The Thinking That Leads (2011)
17. Wells, C.: Sweden Forced to Import Trash from Norway to Create Heat and Electricity in Daily News. New York (2012)
18. Khansari, N., Mostashari, A., Mansouri, M.: Impacting sustainable behavior and planning in smart city. Int. J. Sustain. Land Use Urban Plann. **1**(2), 46–61 (2013)
19. Sapci, O., Considine, T.: The link between environmental attitudes and energy consumption behavior. J. Behav Exp. Econ. **52**, 29–34 (2014)
20. Hobson, K.: Thinking habits into action: the role of knowledge and process in questioning household consumption practices. Local Environ. **8**(1), 95–112 (2003)
21. Opschoor, H., van der Straaten, J.: Sustainable development: an institutional approach. Ecol. Econ. **7**(3), 203–222 (1993)
22. Khansari, N., Mostashari, A., Mansouri, M.: Conceptual modeling of the impact of smart cities on household energy consumption. Procedia Comput. Sci. **28**, 81–86 (2014)
23. Pothitou, M., et al.: A framework for targeting household energy savings through habitual behavioural change. Int. J. Sustain. Energy (ahead-of-print): 1–15 (2014)
24. Khansari, N., A. Mostashari, Mansouri, M.: Impact of information sharing on energy behavior: a system dynamics approach. In: 2014 8th Annual Systems Conference (SysCon) IEEE. IEEE (2014)
25. Kemp, R.: Technology and the transition to environmental sustainability: the problem of technological regime shifts. Futures **26**(10), 1023–1046 (1994)

Diversity Measures for Smart Cities

Ronald R. Yager

Abstract Diversity is a key consideration in the construction of smart cities. With this in mind we introduce a measure of diversity related to the problem of selecting of selecting n objects from a pool of candidates lying in q categories. We introduce the concept of target diversity index (TDI) and describe various agendas for defining it. We look at the problem of trying to select elements to satisfy some desirable criterion that additionally satisfy a requirement of being diverse. We suggest some aggregation methods for combining these multiple criteria.

Keywords Diversity · Social engineering · Multiple criteria decisions · Target diversity index

1 Introduction

Smart Cognitive cities [1–6] has the promise to bring many benefits to our urban environments. However the costs of introducing the required technologies to implement these benefits will be very high and many of the aspects of Smart cities will have to be implemented in a selective way. As our urban environments have become more heterogeneous and social imperatives such as political correctness have gained prominence requirements for diversity and fairness will become an important consideration in many governmental decisions about the allocation of the needed resources for the implementation of Smart Cognitive cities. In anticipation of this, we began to here to provide some tools that can help in addressing concerns about diversity.

Fundamental to considerations of diversity is a partitioning of some population into different categories based upon some demographic feature associated with members of the population. While the issue of choice of categories is important, as it implicitly implies a degree of entitlement based on some demographic, we shall

R.R. Yager (✉)
Machine Intelligence Institute, Iona College, New Rochelle, NY 10801, USA
e-mail: yager@panix.com

© Springer International Publishing Switzerland 2016   197
E. Portmann and M. Finger (eds.), *Towards Cognitive Cities*,
Studies in Systems, Decision and Control 63, DOI 10.1007/978-3-319-33798-2_10

not address this issue, we shall assume the existence of the categories into which the population has been partitioned.

Here we introduce some measures of diversity associated with the problem of selecting n objects from a population so that the various categories that form a partitioning of the population from which these objects are drawn are appropriately represented. After introducing this measure of diversity we look at the multi-objective problem of trying to select elements to satisfy some desirable criterion that additionally satisfy a requirement of being diverse.

2 Diversity Measures in Choosing from a Population

Our goal here is to obtain a formulation that allows us to measure the satisfaction of a criterion of diversity in a selection process. Assume we have a population of objects that fall into q categories and we want to select n elements. Let us denote these categories as C_i for i = 1 to q. Assume we select n_i elements from category C_i. Let us denote this allocation as $A = [n_i, \ldots, n_q]$. We can consider diversity with respect to the ratio $r_i = \frac{n_i}{n}$. The closer all these ratios are to being equal the more we have satisfied a concept of diversity. The situation in which one of the $n_i = n$ and all others are zero can be seen as being the most egregious with respect to satisfying a condition of diversity. Here one of categories, i.e. C_k, has $r_k = 1$ and all other categories have $r_i = 0$. Let us obtain a measure of diversity using this idea. Let us associate with each r_i a value $p_i = \frac{r_i}{\sum_{j=1}^{q} r_j}$. We easily see the p_i have the properties that each $p_i \in [0, 1]$ and there sum is one. Thus the p_i has the nature of a probability distribution. We can measure the satisfaction of the requirement for diversity using an the entropy measure [7–10] associated with this probability distribution obtained from A, $H(A) = -\sum_{i=1}^{q} p_i \ln(p_i)$. Let us first more carefully look at the formulation for p_i. We see that $p_i = \frac{r_i}{\sum_{j=1}^{q} r_j} = \frac{\frac{n_i}{n}}{\sum_{j=1}^{q} \frac{n_j}{n}} = \frac{n_i}{\sum_{j=1}^{q} n_j} = \frac{n_i}{n}$. Here we see that $H(A) = -\sum_{i=1}^{q} p_i \ln(p_i) = -\sum_{i=1}^{q} \frac{n_i}{n} \ln(\frac{n_i}{n})$.

It is well known [10] that this measure takes its largest value when all p_i are the same, $p_i = 1/q$. Here we have $n_i = n/q$. In this case $H(A) = -\sum_{i=1}^{q} \frac{1}{q} \log(\frac{1}{q}) = \log (q)$. This measure takes its smallest value when one of the $p_i = 1$ and all others are zero. In this case $n_k = n$ and all others are zero. In this case the entropy is zero. Here we have taken all the elements from one category and we clearly have the least diversity in our selection. Thus this entropy measure lies between zero and log(q). Using this we can provide a normalized measure that can be used to measure the satisfaction of diversity

$$\text{Div}(A) = \frac{-\sum_{i=1}^{q} p_i \log (p_i)}{\log (q)} = \frac{-\sum_{i=1}^{q} (\frac{n_i}{n})\log (\frac{n_i}{n})}{\log (q)}$$

We see $\text{Div}(A) \in [0, 1]$ and the larger $\text{Div}(A)$ the more the requirement of diversity is satisfied. Thus $\text{Div}(A)$ provides a useful measure of diversity in the case where diversity is captured by a desire for all $\frac{n_i}{n}$ to be equal.

One problem with the above measure is that it doesn't take into account information about the categories from which we choose the n elements. This can be seen as a measure of **pure diversity**. Let us look more carefully at this problem. Assume we have a collection of T objects and we want to select $n \leq T$ objects from this collection. Assume again we assume we have q categories into which the objects in T are divided. Each object falls in one category. If T_i is the number of members in category C_i then $\sum_{i=1}^{q} T_i = T$. Now assume we select n objects from T such that under the allocation A we select $n_i \leq T_i$ elements from category C_i. Here then $\Sigma_i n_i = n$. The question now becomes how can we measure the diversity of this allocation so that we take into account the number of elements in each of the categories. The following formulation can be used.

Here we shall let $r_i = \frac{n_i}{T_i}$ and again our goal for complete diversity is for all the r_i to be equal. Again as in the preceding we shall define $p_i = \frac{r_i}{\sum_{j=1}^{q} r_j}$. It is easily to see that the p_i have the nature of a probability distribution, all the p_i are contained in the unit interval and their sum is one. Using this definition for p_i we again calculate $\text{Div}(A) = \frac{-\sum_{i=1}^{q} p_i \log (p_i)}{\log (q)}$.

We see that if we choose all the elements that are selected from one category C_k, $n_k = n$ then $r_k = \frac{n}{T_k}$ and $r_i = 0$ for $i \neq k$. From this we get $p_k = 1$ and $p_i = 0$ for $i \neq k$ and hence $\text{Div}(A) = 0$. On the other hand if all the $\frac{n_i}{T_i}$ are equal, $\frac{n_i}{T_i} = a$, then all p_i are equal, $p_i = \frac{1}{q}$ then $\text{Div}(A) = 1$. Thus here we see the most diversity occurs for the case of an equal proportional allocation.

A fuller expression of this measure of diversity is provided below

$$\text{Div}(A) = \frac{-\sum_{i=1}^{q} p_i \log (p_i)}{\log (q)} = -(\frac{1}{\log q}) \sum_{i=1}^{q} \frac{(\frac{n_i}{T_i})}{\sum_{j=1}^{q} \frac{n_j}{T_j}} \log(\frac{(\frac{n_i}{T_i})}{\sum_{j=1}^{q} \frac{n_j}{T_j}})$$

$$\text{Div}(A) = -\frac{1}{\log (q)} \frac{1}{\sum_{j=1}^{q} (\frac{n_j}{T_j})} \sum_{i=1}^{q} \frac{n_i}{T_i} (\log(\frac{n_i}{T_i}) - \log(\sum_{i=1}^{q} \frac{n_j}{T_j}))$$

In some instances the diversity allocation associated with a category is best measured with respect to some other population of this category than the number of elements T_i in the set from which we must choose. This may occur for example if there is a belief that there is some bias in the creation set of elements from which we must choose.

Assume again we have a collection of T objects, which we shall refer to as our *choice set*, and we want to select $n \leq T$ objects from this choice set. Assume we have q categories in which the elements are divided, each element falls in one category. Again we let T_i be the number of elements in each category, $\sum_{i=1}^{q} T_i = T$. Here again an allocation A consists of a selection of n_i elements from the ith category, where $n_i \leq T_i$. However, here we have a universal set $S > T$ such that S_i are the number of elements in the universal set that are in the ith category. In this case we shall let $u_i = S_i/S$ be the proportion of elements in S from the ith category. We can now provide a measure of the diversity of the allocation A with respect to S. In this case we shall define $r_i = \frac{n_i}{S_i}$ and again our goal for complete diversity is for all the r_i to be equal. Again as in the preceding we shall define and let our measure of diversity be $\mathrm{Div}\,(A/S) = -\sum_{i=1}^{q} p_i \log p_i$ where $p_i = \dfrac{\frac{n_i}{S_i}}{\sum_{j=1}^{q} \frac{n_j}{S_j}}$. Since $S_i = Su_i$ we can express $p_i = \dfrac{\frac{n_i}{u_i S}}{\sum_{j=1}^{q} \frac{n_j}{u_j S}} = \dfrac{\frac{n_i}{u_i}}{\sum_{j=1}^{q} \frac{n_j}{u_j}}$.

We note that in this case it may be impossible to completely satisfy the diversity condition, $\mathrm{Div}(A/S) = 1$. We see this as follows. Full diversity requires $\frac{n_i}{u_i} = \frac{n_j}{u_j}$ for all i and j. This requires $n_i = \frac{u_i}{u_j} n_j$. If category i has a small number of elements it may be impossible to meet the requirement.

We now want to provide some measure of diversity that takes into account both the choice set population T and the universal set population S.

Let us look at our original formulation with just the choice set. We let $r_i = \frac{n_i}{T_i}$ and then defined

$$p_i = \frac{r_i}{\sum_{j=1}^{q} r_j} = \frac{\frac{n_i}{T_i}}{\sum_{j=1}^{q} \frac{n_j}{T_j}}$$

If we denote $v_i = T_i/T$ then we express

$$p_i = \frac{\frac{n_i}{v_i T}}{\sum_{j=1}^{q} \frac{n_j}{v_j T}} = \frac{\frac{n_i}{v_i}}{\sum_{j=1}^{q} \frac{n_j}{v_j}}$$

Here then if we denote $a_i = \frac{n_i}{v_i}$ then $p_i = \frac{a_i}{\sum_{i=1}^{q} a_i}$.

Similarly we showed for the case of universal population that $p_i = \dfrac{\frac{n_i}{u_i}}{\sum_{j=1}^{q} \frac{n_j}{u_j}}$ and if we let $d_i = \frac{n_i}{u_i}$ then $p_i = \frac{d_i}{\sum_{j=1}^{q} d_j}$.

We now can suggest an approach to combine these and consider the satisfaction of both. We let

$$h_i = \alpha\, a_i + (1 - \alpha)d_i$$

where $\alpha \in [0, 1]$. From this we obtain $p_i = \dfrac{h_i}{\sum_{j=1}^{q} h_j}$. Using this we can measure the diversity of the allocation $A = < n_1, n_2, \ldots n_q >$ where n_i items are selected for the ith category as

$$\text{Div}\,(A) = -\sum_{i=1}^{q} p_i \log\,(p_i)$$

Thus α allows us to balance between using only the choice set population, $\alpha = 1$ and only using the universal population, $\alpha = 0$. We shall refer to α as the measure of locality, in determining diversity. We note that

$$h_i = \alpha \frac{n_i}{v_i} + \frac{(1 - \alpha)n_i}{u_i} = n_i\Big(\frac{\alpha u_i + \bar{\alpha}\, v_i}{v_i u_i}\Big).$$

We note that we can view V as a vector with components v_i and U as a vector with components u_i. Each of these vectors has elements lying in the unit interval and summing to one. We can use $1 - \sum_{j=1}^{q} (u_j - v_j)^2$ to measure the compatibility of the proportions of elements in the choice set T and the global population S.

3   Measuring the Diversity When Objects are in Multiple Categories

In some cases each of the objects in the choice set T rather than being uniquely in one category can partially belong to more than one category. Here again we shall assume q categories. Let $X = \{x_1, \ldots., x_T\}$ indicate the T elements in the choice set . We now associate with each x_j a q-vector $M_j = [m_{j1}, \ldots, m_{jk}, \ldots., m_{jq}]$ where $m_{jk} \geq 0$ and $\sum_{k=1}^{q} m_{jk} = 1$ such that m_{jk} is the degree to which x_j is considered a member of the ith category.

Assume B is a subset of X consisting of n elements. We can associate with B a function called its characterization function, such that $B(x_j) = 1$ if $x_j \in B$ and $B(x_j) = 0$ if $x_j \notin B$. In this case we calculate $n_i = \sum_{x_j \in B} m_{ji}$ as the amount of cate-

gory i in the chosen set B, we see $n_i = \sum_{i=1}^{T} m_{ji} B(x_i)$. We now can get the measure of pure diversity as

$$Div(B) = -\frac{1}{\log q} \sum_{i=1}^{q} \frac{n_i}{n} \log \left(\frac{n_i}{n}\right)$$

When can generalize to this to case where we are interested in determining the diversity of selecting B with respect to overall diversity of the set X. In this case we still calculate $n_i = \sum_{j=1}^{T} m_{ji} B(x_j)$. However here we additional calculate $T_i = \sum_{j=1}^{T} m_{ji}$, it is the total amount of ith category in the set X. Using this we again let $r_i = n_i/T_i$ and $p_i = \frac{r_i}{\sum_{j=1}^{q} r_j}$ and then get

$$Div (B/T) = -\frac{1}{\log q} \sum p_i \log p_i.$$

We can also similarly extend this to case where we use the global set S.

We should note that our original problem, when each element is in exactly one category, is a special case of the preceding. In that case then the vectors M_j are such that one of components has value one and all other components have value zero.

4   Alternative Measures of Diversity

Given a situation in which we have q categories in the preceding we have considered the calculation of the measure of diversity from three perspectives, the pure, with respect to the diversity in the choice set and with respect to the diversity in the universal set. We have shown an underlying uniformity in these three perspectives. In the following we shall use the notation u_i associated with the proportion of the elements of category i in the universe. We note that the other two perspectives can be seen as special cases of this perspective. In the perspective of pure diversity we have a special universe in which each category has the same number of elements. In this case we have $u_i = \frac{1}{q}$ for all i. In the second case we can consider the inverse as the choice set and obtain u_i as the proportion of the elements of category i in the choice set. Thus here then, unless otherwise stated, we shall consider the measure of diversity with respect u_i. We shall refer to u_i as the target diversity index (**TDI**) and refer to the collection of these as U. Here then we have $U = (u_1, u_2, \ldots u_i, . u_q)$.

We assume in the following that B is a selection of n elements from the choice set and n_i is the number of element from category i in B, we have $\sum_{i=1}^{q} n_i = n$. Here we shall denote the allocation of the n elements in B as $A = (n_1, n_2, \ldots n_i, . n_q)$. Here with u_i indicating the target diversity index for the ith category we shall denote $b_i = n_i/u_i$. As discussed attaining diversity is related to the goal of getting all b_i equal. Letting $p_i = \frac{b_i}{\sum_{j=1}^{q} b_j}$ we have suggested calculating the degree of diversity attained by B with respect to the target diversity indices U as

$$\mathrm{Div(B/U)} = -\frac{1}{\log q} \sum_{i=1}^{q} p_i \ln(p_i)$$

We see that when all $n_i/u_i = K$ then all $p_i = \frac{1}{q}$, and the diversity is maximal. When one of the $p_i = 1$ and all other zero the diversity is minimal. Here the measure of diversity is based on the entropy like quantity $-\sum_{i=1}^{q} p_i \ln(p_i)$. Beside these boundary conditions for maximal and minimal values the key feature associated with this measure of diversity is what is called preference for equal division [11]. In [11] we investigated a measure of dispersion associated with the OWA operator which is closely related to the measure of diversity. Here we draw upon the results provided in [11, 12] to suggest some alternative formulation for the measure of diversity.

Again in the following unless otherwise indicated we have $p_i = \frac{b_i}{\sum_{j=1}^{q} b_i}$ where $b_i = n_i/u_i$. Based on the results obtained in [11, 12] one possible formulation for the measure of diversity can be based on the use of the Gini entropy [13] measure $M = -\sum_{i=1}^{q} p_i^2$. Using this entropy measure we see that when all $p_i = 1/q$, the most diverse case, we get $M = -1/q$. At the other extreme when one of the $p_i = 1$ and all others are zero, the least diverse case, we get $M = -1$. From this we can obtain a normalized measure of diversity as

$$\mathrm{Div(A)} = \frac{q}{q-1}[1 - \sum_{j=1}^{q} p_i^2]$$

In [11] the author suggests other functions that can be used to measure entropy.

5   Calculating the Satisfaction of Subsets of Objects to a Criteria

For the development of smart cities we generally want the best people to be employed in significant jobs. As a result of this requirement a task that arises in many situations is the selection of n objects based upon their performance on some test. For example entrance to some elite schools is based on the candidates performance on some standard test, the selection from candidates for some government jobs such as administrators, policemen or fireman is often based on some kind of scoring based on their qualification for the job.

With this in mind we now consider the following problem. Assume we have a collection of T candidate objects, $X = \{x_1, \ldots, x_T\}$. Assume we have an ordered list L of these objects regarding their satisfaction to some criteria. Here we shall allow ties, thus our list is a preference ordering [14]. In this case we can associate with each x_j a number, L_j equal to its position in the list. We note that if two elements are tied for the fifth position then they both get an L value of five and the next element has an L value of seven. Thus L_j is one plus the number of elements ahead of x_j on the list. We are now interested in selecting the n best elements in X satisfying this criteria. Let A be some selection of n elements from T for which we are interested in determining the degree to which selecting these n objects corresponds to the satisfying requirement of selecting the n best. In following we introduce a measure to capture the satisfaction of this requirement. First we associate with each x_j a value b_j such that

$$b_j = 1 \quad \text{if } L_j \leq n$$
$$b_j = 0 \quad \text{if } L_j > n$$

We note that we are taking into account the situation where there are ties at the last position. In particular in this case we get more than n elements with $b_j = 1$.

We next obtain the characteristic function of A. Here $A(x_j) = 1$ if $x_j \in A$ and $A(x_j) = 0$ if $x_j \notin A$, We shall let $a_j = A(x_j)$. Using this we define $e_j = a_j b_j$. We note both a_j and b_j are available for all T elements in X.

Before proceeding we recall the OWA aggregation operation [15, 16]. Let $(g_1, \ldots, g_T)$ be a collection of arguments. Let W be a vector of T weights w_j such that $w_j \in [0, 1]$ and $\sum w_j = 1$. We define the OWA aggregation of the $(g_1, \ldots, g_T)$ under W as

$$OWA_W(g_1, \ldots, g_T) = \sum_{j=1}^{T} w_j g_{ind(j)}$$

where ind(j) is the index of the jth largest of the g_i.

We now can calculate the OWA aggregation of both the b_j and the e_j using a weighting vector W such that $w_j = 1/n$ for $j = 1$ to n and $w_j = 0$ for $j > n$. In particular we obtain $B = OWA_W(b_1, \ldots, b_T)$ and $E = OWA_W(e_1, \ldots, e_r)$. Using these we calculate the degree to which A satisfies the criteria of containing the n best elements from the list L as

$$Sat(L/A) = E/B.$$

We note that if there are r elements in A that are not in the best n these have $b_j = 0$ and then we get

$$Sat(L/A) = \frac{n-r}{n} = 1 - \frac{r}{n}.$$

Here we now extend our interest to the case in which, rather than simply having an ordered list, we have some numeric quantification of how the individual objects in X satisfy the criteria of interest. Here we let $c_j \in [0, 1]$ indicate the degree to which element x_j satisfies the criteria C. Our objective here is again to select the n objects giving us the most satisfaction to this criterion. Here we proceed as follows. We first assign the value c_j to the b_j, thus $b_j = c_j$. Having this we proceed as before. If we select the subset of A of n elements then we let $e_j = a_j b_j$, here a_j is again the characteristic function of A. Using this we calculate $B = OWA_W(b_1, \ldots, b_T)$ and $E = OWA_W(e_1, \ldots, e_n)$ and then calculate our satisfaction as $Sat(C/A) = E/B$. We observe that since $a_i = 0$ if $x_i \neq A$, then E is average the score of the elements in A. We also observe that B is the average of the n largest values of c_j. Thus we see that $Sat(C/A) = \frac{E}{B}$ is the ratio of the average satisfaction in A to largest possible satisfaction by n elements.

We also observe that if A consists of the set of elements with the n largest values for c_j then $e_j = b_j$ and $E = B$ and hence $Sat(C/A) = 1$, this is the largest value for $Sat(C/A)$. We shall denote the set of n elements with the largest for c_j as A:C . Thus A:C gives us the largest satisfaction to our criteria.

We note we any set A of n elements

$$Sat(C/A) = \frac{E}{B} = \frac{B+E-B}{B} = 1 - \frac{(B-E)}{B}$$

where E is the average score of elements in A and B is the average of score of the elements in A:C.

6   Satisfying a Criteria with Diverse Elements

We now turn to our main problem of interest. We have a collection of objects $X = \{x_1, \ldots, x_T\}$. We have some criteria, C, for which $c_j \in [0, 1]$ indicates the degree to which x_j satisfies this criteria. Our object here is to select n objects from X that best satisfy this criterion. However in some situations because of political reasons, such as political correctness, we have the additional requirement that the selected elements should meet some condition of being diverse. In particular here we have q categories and associated with each category is a target diversity index u_i such that $u_i \in [0, 1]$ and $\Sigma_i u_i = 1$.

We can use the formulations developed in the preceding to help in making this decision. Assume A is some subset of n elements from X. Let n_i be the number of elements in A that are in category i. We can calculate the degree to which the condition of diversity is satisfied by A as.

$$Div(A) = -\frac{\sum_{i=1}^{q} p_i \log(p_i)}{\log(q)}$$

where with $d_i = n_i/u_i$ we have $p_i = \frac{d_i}{\sum_{i=1}^{q} d_i}$.

In addition we can calculate the satisfaction of criteria C by the selection of the n element in A as $Sat(C/A) = E/B$ where E and B are as defined earlier.

The determination of the overall satisfaction to the two criteria requires an aggregation of Sat(C/A) and Div(A). We denote

$$\text{Score}(A) = \text{Agg}(\text{Sat}(C/A), \ \text{Div}(A))$$

The form of the Agg function is a reflection the relationship between the two objectives and their respective importances.

In the simplest case we can associate with the satisfaction of C an importance weight $\alpha_C \in [0, 1]$ and with the satisfaction of the diversity condition an importance weight $\alpha_D = 1 - \alpha_C$ and then obtain $\text{Score}(A) = \alpha_C \, \text{Sat}(C/A) + \alpha_D \, \text{Div}(A)$. If both are of the same importance then $\alpha_C = \alpha_D = 0.5$.

In the preceding we have suggested a function, Score(A) to measure the overall satisfaction to the objectives of diversity and some criteria C when selecting the n elements in C. One possible use of this Score is an objective function in some optimization problem. Here our goal is to find the selection A that maximizes the value of Score(A). Often finding the solution of this optimization problem is rather difficult. Another use of this score function is to compare suggested different subsets of n elements. In this case we are essentially obtaining a sub-optimal solution.

An important issue here is the idea of the cost of diversity. As we indicated earlier given the criterion C the maximal satisfaction to this criterion is obtained by selecting the subset consisting of those n elements with the highest value for C, we denoted this set as A:C. In this case Sat(C/A:C) = 1. If the additional requirement of satisfying diversity leads us to select the subset A then we get Sat(C/A). We see that

$$\text{Sat}(C/A) = 1 - \frac{\sum_{j \in A:C} c_j - \sum_{j \in A} c_j}{\sum_{j \in A:C} c_j}$$

Letting $V = \sum_{j \in A:C} c_j$ and $V_A = \sum_{j \in A} c_j$ then $\text{Sat}(C/A) = 1 - \frac{(V - V_A)}{V}$. We calculate the loss of satisfaction to C due to diversity as

$$\text{Sat}(C/A:C) - \text{Sat}(C/A) = \frac{(V - V_A)}{V}$$

We denote this as the **COD, cost of diversity**.

7   Other Formulations for Overall Satisfaction

In the preceding we suggested using a weighted average of the satisfactions to diversity and C by the selection A. Here the weights where the importances of each of these two objectives. Other formulations for aggregation of the satisfactions of the diversity and C are possible [17]. Here we briefly describe some others.

Another formulation for Score(A) can be obtained if we consider a prioritize relationship between diversity and criteria C. We briefly describe this approach we refer the reader to [18, 19] for a fuller discussion of this type of aggregation model. Consider the case where diversity has priority over the criteria C. In this case we require that any selection A satisfy diversity, satisfaction to C can't compensate for a lack of diversity. Only if a solution A satisfies diversity can we consider its satisfaction to C. In this case we calculate

$$\text{Score}(A) = w_1 \, \text{Div}(A) + w_2 \, \text{Sat}(C/A)$$

where

$$w_1 = \frac{1}{1 + \text{Div}(A)}$$

$$w_2 = \frac{\text{Div}(A)}{1 + \text{Div}(A)}$$

We see if the satisfaction of diversity, Div(A), is small then w_2 is small and hence even a high value for Sat(C/A) can't compensate. This formula is very closely related to lexicographic decision-making [20]. We note that the amount of satisfaction to C by A can only significantly contribute to the score if the diversity condition is satisfied by A.

In the case where criteria C has priority over diversity then

$$\text{Score}(A) = w_1 \, \text{Sat}(C/A) + w_2 \, \text{Div}(A)$$

where

$$w_1 = \frac{1}{1 + \text{Sat}(C/A)}$$

$$w_2 = \frac{\text{Sat}(C/A)}{1 + \text{Sat}(C/A)}$$

Another consideration that can be used to help in the construction of score is what has been called the organization of the criteria. Here we let $\lambda \in [0, 1]$ be a parameter such that if $\lambda = 1$ we need to satisfy both of the criteria, diversity and C, while if $\lambda = 0$ we require only the satisfaction any of the two. Thus $\lambda = 0$ indicates that we need diversity **or** the satisfaction of C. While $\lambda = 1$ indicates that we need to satisfy both diversity **and** criterion C. In point of fact $\lambda = 1$ indicates we want **all** the criteria while $\lambda = 0$ indicates we need only satisfy **any** of the criteria.

With this type of relationship between the two the objectives, satisfaction of C and diversity, we can obtain a formulation for Score(A) using a special form of the OWA operator with two arguments [21]. We describe this formulation in the following. Here we let

$$m_1 = \text{Max}[\text{Sat}(C/A), \ \text{Div}(A)]$$
$$m_2 = \text{Min}[\text{Sat}(C/A), \ \text{Div}(A)]$$

With α_C and $\alpha_D = 1-\alpha_C$ as defined previously as the importances of objectives we let

$$\alpha = \alpha_C \quad \text{if } m_1 = \text{Sat}(C/A)$$
$$\alpha = \alpha_D \quad \text{if } m_1 = \text{Div}(A)$$

Using this we obtain

$$\text{Score}(A) = m_1 \alpha^{\tan\left(\lambda\frac{\pi}{2}\right)} + m_2\left(1 - \alpha^{\tan\left(\lambda\frac{\pi}{2}\right)}\right)$$

We can recall tan is the trigonometric function with $\tan(0) = 0$, $\tan(\pi/4) = 1$ and $\tan(\pi/2) = \infty$.

We see that

$$\text{If } \lambda = 1 \text{ then Score}(A) = \text{Min}(\text{Sat}(C/A), \ \text{Div}(A))$$
$$\text{If } \lambda = 0 \text{ then Score}(A) = \text{Max}(\text{Sat}(C/A), \ \text{Div}(A))$$

On the other if $\lambda = 0.5$, then $\tan(\lambda\pi/2) = 1$ and we get $\text{Score}(A) = \alpha_C \text{Sat}(C/A) + \alpha_D \text{Div}(A)$

In the preceding we included the requirement of diversity using our formula Div (A). It is possible to include the requirement for diversity using the idea of linguistic values and fuzzy sets [22, 23]. Let us denote $U = \text{Div}(A)$. We see that U can be viewed as a variable that takes values in the unit interval depending on A. One can associate with this variable, linguistic values such as high, medium and low. Here we can define a concept such as high as a fuzzy subset of the unit interval, HIGH, such that for each value $r \in [0, 1]$, HIGH(r) indicates the degree to which r satisfies the concept of high diversity. Various other linguistic concepts can be defined in a similar way. Using these structures we can replace our requirement for diversity in the score function by some linguistically expressed requirement for a level of diversity. Here then if F is a fuzzy subset describing our desired level of diversity then for any selection A, F(Div(A)) would indicate our degree of satisfaction to F. Using this idea we can in our formulation for Score(A) replace Div(A) by F(Div(A)). This allows more sophisticated description of our requirement for the diversity condition.

8 Conclusion

We provided a measure of diversity related to the problem of selecting of selecting n objects from a pool of candidates lying in q categories. We introduced the concept of target diversity index (TDI) and described various agendas for defining it. We looked at the problem of trying to select elements to satisfy some desirable criterion that additionally satisfy a requirement of being diverse. We introduced a quantification of the cost of diversity. We suggested some aggregation methods for combining these multiple criteria.

References

1. Deakin, M., Al Waer H.: From Intelligent to Smart Cities, Routledge, New York (2012)
2. Dameri, R.P., Rosenthal-Sabroux, C.: Smart City: How to Create Public and Economic Value with High Technology in Urban Space (Progress in IS). Springer, Heidelberg (2014)
3. Araya, D.: Smart Cities as Democratic Ecologies. Palgrave Macmillan, New York (2015)
4. Komninos, N.: The Age of Intelligent Cities: Smart Environments and Innovation-for-all Strategies. Routledge, New York (2015)
5. Weston, M.: Smart Cities' Will Know Everything About You. In: The Wall Street Journal, (2015)
6. Stimmel, C.L.: Building smart Cities: Analytics, ICT, and Design Thinking. CRC Press, Boca Raton (2016)
7. Shannon, C.L.: The mathematical theory of communication. Bell Syst. Tech. J. **27**(379–423), 623–656 (1948)
8. Aczel, J., Daroczy, Z.: On Measures of Information and their Characterizations. Academic, New York (1975)
9. Buck, B.: Maximum Entropy in Action: A Collection of Expository Essays. Oxford University Press, NY (1991)
10. Klir, G.J.: Uncertainty and Information. John Wiley & Sons, New York (2006)
11. Yager, R.R.: On the dispersion measure of OWA operators. Inf. Sci. **179**, 3908–3919 (2009)
12. Fuller, R., Majlender, P.: On obtaining minimal variability OWA weights. Fuzzy Sets Syst. **136**, 203–215 (2003)
13. Gini, C.: "Variabilità e mutabilità," Reprinted in Memorie di Metodologica Statistica. In: E. Pizetti and T. Salvemini, (eds.) Libreria Eredi Virgilio Veschi, Rome (1955)
14. Roubens, M., Vincke, P.: Preference Modeling. Springer-Verlag, Berlin (1989)
15. Yager, R.R.: On ordered weighted averaging aggregation operators in multi-criteria decision making. IEEE Trans. Syst. Man Cybern. **18**, 183–190 (1988)
16. Yager, R.R., Kacprzyk, J.: The Ordered Weighted Averaging Operators: Theory and Applications. Kluwer, Norwell, MA (1997)
17. Beliakov, G., Pradera, A., Calvo, T.: Aggregation Functions: A Guide for Practitioners. Springer, Heidelberg (2007)
18. Yager, R.R.: Prioritized aggregation operators. Int. J. Approximate Reasoning **48**, 263–274 (2008)
19. Yager, R.R., Reformat, M., Ly, C.: Web PET: An on-line tool for lexicographically choosing purchases by combining user preferences and customer reviews, IEEE Intelligent Systems, (To Appear)
20. Isermann, H.: Lexicographic goal programming: the linear case. In: Grauer, M., Lewandowski, A., Wierzbicki, A.P. (eds.) Multiobjective and Stochastic Optimization. International Institute for Applied Systems Analysis, Laxenberg, Austria, pp. 65–78, (1982)
21. Yager, R.R.: Quantifier guided aggregation using OWA operators. Int J Intell Syst **11**, 49–73 (1996)
22. Zadeh, L.A.: Outline of a new approach to the analysis of complex systems and decision processes. IEEE Trans. Syst. Man Cybern. **3**, 28–44 (1973)
23. Zadeh, L.A.: Fuzzy logic = computing with words. IEEE Trans. Fuzzy Syst. **4**, 103–111 (1996)

Glossary

Aggregated Information refers to information elements whose values have been generated by performing a calculation across all individual units as a whole. While uncovering new treatment strategies, medical researchers may use aggregated patient data—e.g., a certain percentage of patients taking a particular combination of drugs who experienced adverse side effects—but would not have the means to connect this data to a specific individual

Citizen Communication The term Citizen communication is usually referred to all the human-to-human and human-to-group communication whose aim is to deal with collective sharing of opinions, feelings and resources in metropolitan and rural areas. Improvement of Citizen communication through computational intelligence can bring an increase in Citizen participation and contribute in the mitigation of a general sense of apathy that has pervaded social life in the recent history of civilization

Cognitive City A cognitive city leverages technology to sense, perceive, and respond to changes in the environment and can therefore improve a system's performance by increasing its adaptive capacity

Cognitive Computing enhances computing with cognition. It has various topics encompassing connectivism, computational thinking as well as the intelligence amplification loop. Cognitive computing systems are able to (automatically) learn and evolve based on data and experience by detecting patterns and deriving meaningful information from various sources (e.g., texts, images). Cognitive computing attempts to improve computer systems (i.e., make them more intelligent) to make it possible to compute with imprecise and complex data and to address reality. Cognitive systems are able to extend human intelligence and to enhance the collaboration between humans and computer systems. They learn and evolve based on data and experience

Cognitive Systems have the capability to learn from domain related data by building models. These models can test, train and generate hypotheses to answer problem statements

© Springer International Publishing Switzerland 2016
E. Portmann and M. Finger (eds.), *Towards Cognitive Cities*,
Studies in Systems, Decision and Control 63, DOI 10.1007/978-3-319-33798-2

Complex Adaptive Systems have a large number of components (i.e., subparts) that interact with each other. They are able to learn and adapt to new situations, resulting from these interactions. These systems are dynamic networks of interactions

De-identified Information are records that have had sufficient personal information removed or obscured in some manner such that the remaining information does not identify an individual, either directly or indirectly and there is no reasonable basis to believe that the information may be used, either alone or with other information, to identify an individual. Proper protocols must be used to remove or modify direct, indirect, and quasi-identifiers; de-identification must be done in conjunction with re-identification risk management procedures

Digital Personal Assistant A digital personal assistant acts as a computerized version of a human personal assistant. It performs tasks similar to a human assistant and is able to perform decision making on its own behalf. The digital personal assistant accesses the data made available and performs them based on the interests and requirements of the given user

e-Collaboration refers to a range of measures for decentralized computer-based collaboration between temporally or spatially separate teams and groups

e-Community Also known as a virtual community, an e-Community is a social network of individuals who interact through specific social media, potentially crossing geographical and political boundaries in order to pursue mutual interests or goals

e-Democracy By electronic democracy, or e-Democracy, we understand the support and enhancement of civil rights and duties in the Information and Knowledge Society. It can be executed with the help of information and communication technologies, independent of time and place

e-Empowerment is the highest level of participation that can provide citizens the opportunity to express their will on various initiatives and projects

e-Government refers to the use of information technologies and knowledge in the internal processes of government and the delivery of products and services of the state to citizens and industry

e-Health The term e-Health describe the use of information technology in medicine for communications between patients and professionals and to exchange physiologic data in order to improve access to healthcare services with economic benefits. Anonymized data collections can be also analyzed in public health for disease surveillance and prevention

Emerging country In an emerging country, the population has a lower standard of living in terms of economic and social conditions (e.g., poor supply situation with food and consumer goods, restrictions in health care, inadequate infrastructure, air and water pollution), compared to other countries

Emoji are pictograms representing a number of concrete and abstract concepts such as faces, vehicles, buildings, animals, plants, emotions, feelings and activities. Available on smartphones and in chats, emoji are very popular in Japan, where they were conceived in 1999, and worldwide. The appearance of emoji can vary significantly in shape, and emoji do not have to look the same on all devices: according to the Unicode Consortium any pictorial representation based on the name is considered an acceptable rendition. Emoji are a compact and quick way to express emotions in a social context mediated by computer technology

e-Participation is the generally accepted term referring to "ICT—supported participation in processes involved in government and governance."

Fuzzy Cognitive Maps (FCMs) FCMs are traditional cognitive maps extended through fuzzy logic consisting of nodes (e.g., concepts) and weighted edges (called weights). The nodes and weighted edges are connected constructing a fuzzy weighted digraph with feedbacks. The connections describe the causal relationships between the nodes and weighted edges. A FCM is a mathematical tool for modeling complex systems built by directed graphs with concepts (e.g., policies, events, and/or domains) as nodes and causalities as edges. The causalities have values in the interval between -1 and 1

Fuzzy Logic is based on fuzzy set theory. Thus, fuzzy logic augments bivalent truth by partial truth, where a proposition can be anything between false or true (i.e., it can take on any degree of truth) and enhances the traditional binary distinction between 0 and 1 with the gradual membership function allowing values between 0 and 1. Linguistic variables can be used as labels for the membership degree (e.g., weak, high.)

Information Communication Technology (ICT) ICT is the hardware and software infrastructure that enables communication through local and global voice, video, and data interconnectivity

Informational Self-determination This term (reflecting the fact that the individual should determine the fate of his/her personal information) is based upon the German Federal Constitutional Court ruling that "…in the context of modern day processing, the protection of the individual against unlimited collection, storage, use and disclosure of his/her personal data is encompassed by the general personal rights of the [German Constitution]. This basic right warrants the capacity of the individual to determine in principle the disclosure and use of his/her personal data. Limitations to this informational self-determination are allowed only in cases of overriding public interest."

Living City A Living City actively enhances the quality of life for its citizens using technology as an enabler. It combines sustainable urban development with urban technology and intelligence

Metaheuristics is a general term indicating a class of algorithms that are designed to provide solutions to problems that are solvable in theory but not in practice, through the use of methodologies that approximate an exact solution and take inspiration from the cognitive domain and the inner working of physical or biological complex systems. Examples of metaheuristics are Simulated Annealing, Tabu Search, Local Search, Variable Neighborhood Search, Ant colony/Particle Swarms, Evolutionary Computation, and Genetic Algorithms

Paper Prototype is a low technology form of prototyping. It allows the simulation of a system's or products most important functions through user testing. Paper prototypes are mainly hand drawn or supported by a software (e.g., Microsoft PowerPoint). Therefore, ideas and designs can rapidly be explored and tested with users

Privacy by Design (PbD) PbD is an internationally recognized framework that consists of 7 Foundational Principles. Developed by Dr. Ann Cavoukian, she had it unanimously passed in 2010 as an "essential component of fundamental privacy protection" by the Assembly of International Privacy Commissioners and Data Protection Authorities. The principles build upon established Fair Information Practice Principles (FIPPs), but raise the bar for privacy and data protection by promoting enhanced accountability and trust through: proactive leadership and goal-setting; adopting a positive-sum, win/win framework; systematic and verifiable implementation methods; and practical and demonstrable outcomes

Privacy enhancing technologies (PETs) PETs refer to a coherent system of information and communication technologies (ICTs) that incorporates privacy-protective principles into technical specifications in order to strengthen privacy by preventing the unnecessary or unlawful collection, use and disclosure of personal data, or by offering tools to enhance an individual's control over his/her personal data, without losing the functionality of the information system. PETs-incorporated systems use such technologies as Identity Protectors to divide systems into identity, pseudo-identity and anonymity domains. PETs are complementary to other instruments with which they must work in order to provide a highly robust form of privacy protection

Privacy Impact Assessment (PIA) At a minimum, a PIA is a tool designed to promote 'privacy by design', better information to individuals, as well as transparency and dialogue with competent authorities. It is also a process which should begin at the earliest possible stages, when there are still opportunities to influence the outcome of a project. It assists with identifying and assessing privacy risks, as well as possible strategies for mitigating those risks throughout the development life cycle of a program, technology or system. The process should continue well after the initial deployment of the project

Restriction-centered Theory Unlike traditional logical systems, RCT can vary over the range from 0 to 1 between true and false. This allows to address the imprecision inherent in language by introducing restrictions

Sentiment Analysis is an umbrella term for a set of data mining techniques that use computational linguistics to detect and extract people's opinions from text written in natural language. The rapid growth of the field coincide with the spread of social media: business companies can now gather public comments about their product from all over the world and make decisions according feedback from consumers

Slum is a bedraggled part of a city, where usually very poor people live. A slum is characterized with buildings that are in a bad state and have inadequate infrastructure

Smart City A smart city leverages information communication technologies such as satellites, wired and wireless technology, sensors, and other forms of data collection devices to help citizens, businesses, and governments to make data-driven, intelligent decisions to address environmental, economic, and social issues

Smart infrastructure embeds information communication technology to provide data, which is used as evidence for informed decision-making

Soft Computing (SC) SC is tolerant of imprecision, uncertainty, vagueness and partial truth. There exist several methods (e.g., fuzzy set theory, granular computing, computing with words) that enable computation with information described in natural language (i.e., imprecise data)

Stakeholder Management is a key factor in every business process or project involving more than one participant. It is about managing the expectations, interests, opinions and needs of the various involved stakeholder. Effective Stakeholder Management creates positive relationships with stakeholders. Stakeholder management is a process and has to be planned and guided

Transdisciplinary Research is a kind of team science. In other words, a collaboration of researchers from different disciplines working jointly, by exchanging information, sharing resources and integrating disciplines to achieve a common scientific goal (e.g., create new innovations) to address a common problem